SpringerBriefs in Applied Sciences and Technology

SpringerBriefs present concise summaries of cutting-edge research and practical applications across a wide spectrum of fields. Featuring compact volumes of 50 to 125 pages, the series covers a range of content from professional to academic.

Typical publications can be:

- A timely report of state-of-the art methods
- An introduction to or a manual for the application of mathematical or computer techniques
- A bridge between new research results, as published in journal articles
- A snapshot of a hot or emerging topic
- An in-depth case study
- A presentation of core concepts that students must understand in order to make independent contributions

SpringerBriefs are characterized by fast, global electronic dissemination, standard publishing contracts, standardized manuscript preparation and formatting guidelines, and expedited production schedules.

On the one hand, **SpringerBriefs in Applied Sciences and Technology** are devoted to the publication of fundamentals and applications within the different classical engineering disciplines as well as in interdisciplinary fields that recently emerged between these areas. On the other hand, as the boundary separating fundamental research and applied technology is more and more dissolving, this series is particularly open to trans-disciplinary topics between fundamental science and engineering.

Indexed by EI-Compendex, SCOPUS and Springerlink.

Kiran Kumar Poloju · Kota Srinivasu

Geopolymer Concrete

Principles, Characteristics, Testing, and Applications

 Springer

Kiran Kumar Poloju
Department of Civil
and Mechanical Engineering
Middle East College
Muscat, Oman

Kota Srinivasu
NRI Institute of Technology
Guntur, Andhra Pradesh, India

ISSN 2191-530X   ISSN 2191-5318 (electronic)
SpringerBriefs in Applied Sciences and Technology
ISBN 978-981-96-2478-2   ISBN 978-981-96-2479-9 (eBook)
https://doi.org/10.1007/978-981-96-2479-9

This Springer imprint is published by the registered company Springer Nature Singapore Pte Ltd.
The registered company address is: 152 Beach Road, #21-01/04 Gateway East, Singapore 189721,
Singapore

If disposing of this product, please recycle the paper.

Acknowledgements

I would like to dedicate this book to my parents (Poloju Nadham and Poloju Prabhavathi), my wife (D. Sounica) and children (Poloju Manomayi Sri and Poloju Akiranandan) and my brothers for their exceptional motivation and moral support to complete this book. I wish to express my sincere gratitude to my co-author and guide Prof. Kota Srinivasu for his esteemed support, my colleagues form the Department of Civil and Mechanical Engineering, and Management of the Middle East College.

About This Book

This textbook focuses on geopolymer concrete, their principles, characteristics, testing, and applications, which is a novel and sustainable replacement for Portland cement concrete. The concrete is made using a mix incorporating industrial byproducts like fly ash, blast furnace slag, and alkaline activators. Geopolymer concrete is eco-friendly, uses waste materials as raw materials, promotes a lighter carbon footprint, and conserves resources. As a fact, geopolymer concrete may fall into the green category, a very eco-friendly attribute of the production process is that it tends to become hardened at room temperature or with negligible heat, leading to far lower energy consumption and emissions. The fundamental idea behind geopolymer chemistry is that an aluminosilicate substance dissolves in a very alkaline solution, and then silica and alumina condense and reorganize into a three-dimensional structure. Consequently, critical observation of geopolymer concrete is necessary. It also provides a more sustainable alternative to Portland cement concrete, the use of which has serious environmental impacts such as high CO_2 emissions, resource consumption, and high energy demand. Moreover, geopolymer concrete has higher durability and chemical resistivity compared to the conventional concrete in civil engineering field, hence it can be suitable for many construction activities. In a nutshell, GPC utilizes waste generated in the industry to produce geopolymer concrete, allowing for reduced waste and conserving natural resources in the context of a sustainable construction sector as an economy.

Contents

About the Authors

Dr. Kiran Kumar Poloju obtained his Ph.D. in civil engineering from Acharya Nagarjuna University, India, and a Master's degree in Structures from the National Institute of Technology, Warangal, India. He is a Microsoft and Google Certified Educator with 13 years of teaching experience in reputable institutions in India and Oman. Dr. Poloju has also been recognized with Fellowship status for meeting the criteria outlined in the UK Professional Standards Framework for Teaching and Learning Support in Higher Education. He has held various positions in the field of civil engineering, including lecturer, senior lecturer, assistant professor, and head of the civil engineering department. Dr. Poloju has an impressive track record of research publications, including textbooks, book chapters, journals, patents, and conference proceedings. He has also participated in various conferences and engaged in numerous funded projects. He holds professional certifications as a chartered engineer and is a life member of various professional bodies like IEI, ICE, IIBE, etc. Dr. Poloju has delivered numerous lectures in the field of civil engineering and has received several awards and accolades for his work. He is expertise person in the implementation of flipped teaching methodology and the use of various e-tools in teaching and learning. His ongoing research and funded projects suggest the potential for continued contributions to addressing challenges in infrastructure resilience and sustainable development. Overall, Dr. Poloju's multifaceted background positions him to make meaningful impacts in both academic and research aspects of civil engineering.

Dr. Kiran Kumar Poloju, Senior Lecturer, Department of Civil and Mechanical Engineering, Middle East College, Muscat, Oman. e-mail: kpoloju@mec.edu.om

Dr. Kota Srinivasu obtained his Ph.D. in civil engineering from the National Institute of Technology, Warangal, India, and his master's degree from the National Institute of Technology, Calicut, India. He served as Head of Civil Engineering and Serving as the Principal of reputed Engineering Colleges since 2006. His academic tenure spans around 40 years, during which he has made substantial contributions to teaching, research, and administration. This extensive period has been distributed across various educational institutions, enabling him to acquire diverse perspectives and expertise in the academic domain. Furthermore, he has supervised 6 Ph.D. students to date and a few more are in progress, secured significant funding from various organizations, and obtained numerous professional certifications. While his extensive academic background undoubtedly provides him with a wealth of knowledge and expertise in educational matters, Dr. Srinivasu's academic experience suggests that he has developed a comprehensive understanding of educational systems, pedagogical approaches, and administrative practices. His long-standing commitment to academia indicates that he has observed and potentially contributed to significant developments in the educational landscape over more than four decades. He visited Stanford University, Google, and Microsoft under Stanford University Innovation Fellows.

Prof. Kota Srinivasu, Principal, NRI Institute of Technology, Perecherla, Guntur, Andhra Pradesh, India.

Chapter 1
Introduction to Geopolymer Concrete

1.1 General

The building industry uses concrete as a material. Concrete is formed when water and cement are mixed with aggregates. Therefore, concrete is usually called "manmade stone," as Badea states. When hydrated, the cement, water, and aggregate become solid. Therefore, energy consumption is the most critical environmental aspect of cement and concrete production. NRMCA (2012) discussed two mechanisms that result in CO_2 generation during cement manufacturing as fossil fuels must be burned.

Heating calcium carbonate releases carbon dioxide and produces calcium oxide. Approximately 90% of the concrete's weight consists of aggregates. Concrete is transported to and from the building site, and at each step in the process, a small amount of CO_2 is emitted through the gravel mining process, stone crushing, and CO_2 production. One cubic meter of concrete uses 10% cement, and one ton of cement releases 9 tons of CO_2. Many studies have been conducted to reduce CO_2 emissions by adding cementitious materials such as fly ash and GGBS. With infrastructure enhancement, cement demand will only increase. Every year, cement manufacturing has increased by 3%. The global impact of it [according to Mehta (2001)] is that the 1.6 billion tons of cement produced around the world emits 7% of all carbon dioxide into the atmosphere. An alternative and sustainable approach is required to mitigate the negative environmental consequences of the building industry. However, there are several cementitious materials such as hypo sludge, ceramic powder, and rice husk ash (Poloju and Srinivasu 2022a, b). Cement manufacturing must be completed to reduce global warming. This can be achieved by using alkali binders with excellent technical merits. GGBS and fly ash are the most used industrial wastes for manufacturing GPC.

K. K. Poloju and K. Srinivasu, *Geopolymer Concrete*,
SpringerBriefs in Applied Sciences and Technology,
https://doi.org/10.1007/978-981-96-2479-9_1

1.2 Background of GPC

The term "geopolymer" was first coined by French scientist Professor Davidovits in 1978 to describe the chemical group of zeolites. The raw materials and alkaline activators are the main components of geopolymers. The aluminum and silica in the binder are activated by an alkaline solution into a gel, which is the strength-yielding ingredient of the GPC in the geopolymer. Large amounts of natural resources in the form of calcareous and siliceous rocks needed for ordinary Portland cement manufacturing are in short supply, driving the search for environmentally friendly and long-lasting alternatives. Therefore, cement alternatives in environmentally friendly concrete that satisfy mechanical, and durability requirements are necessary.

In addition, this material is preferable to conventional cement-based concrete. Several disposal challenges have been created by the thermal power generation of large amounts of fly ash. While this has been attempted by the government, non-governmental organizations, and research and development organizations, fly ash utilization is only around half. However, Bilodeau showed that less than 15% of fly ash is used in construction materials, such as concrete and building blocks. Over the past few years, this proportion has remained constant. The remaining 85% goes to landfills. According to some sources, fly ash is used in the United States at about 32%, and approximately 40% in China. This quantity was similar to that in India. In recent years, geopolymer concrete has become an innovative and ecologically friendly alternative to regular Portland cement concrete. The chemical interaction between industrial wastes (e.g., fly ash or blast furnace slag) and alkaline activators generates this innovative substance, which is formed by a high-performance and durable binder. Geopolymer concrete is attracting interest due to its capacity to resolve many of the problems associated with conventional concrete manufacturing, including large carbon dioxide emissions, resource depletion, and energy demand. The environmental sustainability of geopolymer concrete is one of its greatest benefits. Concrete manufacturing considerably minimizes the carbon footprint by using industrial waste materials as the main components.

First, monitoring of fly ash, a byproduct of coal combustion (or blast furnace slag, a waste product of steel manufacturing), diverts these materials from landfills, and second, it alleviates the demand for raw material extraction. This technique is in line with circular economy principles that promote resource efficiency and waste reduction in the building sector. The manufacturing process makes geopolymer concrete environmentally friendly and even more advantageous. Unlike Portland cement, which requires high-temperature kilns for clinker formation, geopolymer concrete can be cured at room temperature or with a small heat input, thereby reducing energy use and associated greenhouse gas emissions. Research has demonstrated that geopolymer concrete reduces the environmental impact of the building industry by up to 80% compared with traditional concrete, helping to reduce carbon dioxide emissions. In addition, geopolymer concrete presents superior mechanical qualities and enhanced resistance to unfavorable climatic conditions, making the material a preferred choice for a broad range of building applications. Geopolymers exhibit a unique chemical structure that provides outstanding compressive strength, which typically exceeds that of conventional concrete. A high strength-to-weight ratio makes it possible to create more cost-effective and lightweight structures, ideally reducing material utilization and transportation costs. In addition, geopolymer concrete is quite durable in difficult environments. Owing to their resistance to acid attack, sulfate attack, and alkali-silica reaction, which are prominent degradation processes in traditional construction concretes, they are more resistant. This added durability also means a longer service life and lower maintenance costs, which contribute to the material's sustainability credentials. In addition, geopolymer concrete is highly resistant to fires. Unlike ordinary concrete, geopolymer concrete does not spall or lose its strength at high temperatures. Its insulating feature makes it ideal for use in locations exposed to fire and places with strict fire safety standards. Geopolymer concrete provides these beneficial qualities, making it a potential choice for many building applications. These include infrastructure projects, such as bridges, tunnels, and pavements that require high durability and reasonable long-term performance. The exceptional strength and fire resistance of geopolymer concrete make it suitable for use in high-rise structures to improve structural performance and increase safety as a real possibility. The material will be resistant to chloride intrusion and corrosion, benefiting marine constructions that are potentially exposed to extreme salty conditions. With advancements in this research scope, geopolymer concrete will play a significant role in future of sustainable buildings and infrastructure development. Mix designs are being optimized to improve curing techniques and standardize testing protocols to foster the widespread adoption of this material. Researchers have also explored the possibility of using other precursors and activators to make geopolymer concretes more sustainable and performant. The interest of academics and industry is driving the advancement of geopolymer technologies. For instance, precast geopolymer products, self-compacting geopolymer concrete, and geopolymer-based repair solutions have been developed. These developments broaden and extend the spectrum of applications of geopolymer concrete and increase its viability in real-world building projects. However, geopolymer concrete has not been widely used. Assured consistent performance is achieved through quality control procedures, suitable design rules, and standards are developed, and long-term durability and safety questions are resolved.

The solution to these issues will require further research and collaboration between academics and industry, as well as ensuring supporting regulations so that sustainable construction materials can be used everywhere. Geopolymer concrete is finally an ecological alternative to the already limited existing (raw) materials. The material also has good mechanical characteristics, is environmentally friendly, and adaptable, making it an attractive alternative to current concrete. Geopolymer concrete acts as an opportunity to reform the future of sustainable building construction in the move for the building industry, and efforts to mitigate its effect on the environment and contribute to sustainability. The novel binding substance of Soviet geopolymers has several technical uses. These geopolymer compounds function as cement binders rather than as Portland cement. Geopolymers are a new technology that converts a byproduct of an industry into a profitable product while solving many environmental issues. The study was scheduled only for water, but if substituted by an alkaline solution, further economic studies are needed. Fly ash and GGBS are available in raw form in places where thermal power plants are operational, especially in India. The growth in infrastructure construction in the last few decades has made concrete a critical building material, owing to the price of concrete becoming comparatively expensive relative to many construction materials, as well as its durability.

1.3 History of GPC

Since its introduction in the late 1970s, geopolymer concrete has evolved significantly, indicative of a sea change in how construction materials are perceived and developed. In the 1990s experts realized that it could offer an immense solution to the serious environmental problems associated with traditional Portland cement manufacturing. As such research and knowledge grew, geopolymer concrete was recognized not only as a more sustainable option but also as a superior one in terms of durability and resistance to chemical attack. The building sector was of great interest in this realization, particularly where there were ample industrial byproducts available for geopolymer manufacturing. The use of fly ash as a partial cement substitute, owing to its cementitious characteristics of silica and calcium, has been the subject of numerous investigations. Thus, fly ash has been employed as a binding agent to form geopolymer concretes. Since 800 BC, human civilization has known how to make cementitious materials. Davidovits stated that the trick was to dissolve rocks in a cement paste and then use the paste to aggregate and sand together. It is also used for making large stone slabs and sculptures. This process was also used to set the large stone blocks required to build the Pharaoh's Pyramid at Cheops. Unfortunately, there are people who have not really adopted this concept. The major components of hydrated cement paste are CSH, $Ca(OH)_2$, and ettringite. According to Hardjito et al., approximately 60% of hydrated cement consists of CSH. Davidovits was the first to convert industrial byproducts into binding materials using alkaline solutions. Wang et al. (1995) later conducted several studies on GPC characteristics. The results show that the geopolymer concrete made from fly ash is very strong. Based on the information reported by Puertas et al. (2000), a strength of approximately 50 MPa was obtained at high temperature in the presence of a greater alkaline solution concentration. However, fly ash-based geopolymers, though better performing, have some drawbacks, such as rapid setting, limited workability, curing regime, and especially outdoor strength. Additionally, geopolymers based on fly ash require high temperatures (60–90 °C) to achieve early stability. The specimens were tested in the laboratory under heat-curing conditions. Nevertheless, such an environment cannot be developed for large-scale projects in real-world applications. To compensate for these drawbacks, we built paths using alkali-activated slag. After alkali-activated slag, Pradip and Hardjito noted concerns over setting times and workability in their applications. Fly ash-based geopolymers cured with a naphthalene-based superplasticizer showed improved workability. Rangan et al. (2006) suggested mixing quantities so everything could be done as a GPC as an OPC. The alkaline solution concentration and AL/FA ratio were used to measure the mix proportions. Because of the current need for high-performance concrete, this study focused on fly ash and GGBS. Research indicates that geopolymer concrete prepared from fly ash and GGBS can substitute for high-performance conventional concrete. As stated by Manjunatha et al. (2014), there is a scarcity of adequate designs and suggestions for GPC using GGBS. Therefore, the current study focuses on the binder type, binder content, alkaline/binder ratio, and appropriate curing and acceptable conditions. The evolution of geopolymer concrete has been propelled by continuous improvements in several areas, such as mix design improvements, better curing processes, and performance-enhancing strategies. Although ordinary fly ash has been extensively studied as a precursor, researchers have actively studied other precursors, such as metakaolin, blast furnace slag, and rice husk ash. As a result, the field of geopolymers has expanded significantly in its raw material options, increasing its versatility and adaptability to both geographical settings and waste streams. In addition, the development of activator solutions and integration of novel admixtures have made significant progress. The workability and setting properties of geopolymer concrete have been significantly improved by these advancements, making geopolymer concrete more appropriate for large-scale use in a variety of building projects. Better rheological characteristics and controlled setting periods have overcome many of the early hurdles in the practical use of geopolymer technology. Over the past few years, considerable research has been conducted to better understand and optimize the microstructural characteristics of geopolymer concretes. X-ray diffraction (XRD), scanning electron microscopy (SEM), and nuclear magnetic resonance (NMR) spectroscopy have been used for advanced characterization to understand the complex chemical mechanisms and structure formation processes that exist in geopolymer systems. This knowledge has allowed researchers to optimize mix formulations and curing conditions to achieve the desired mechanical and durability characteristics. The environmentally beneficial properties of geopolymer concrete are increasingly apparent from life cycle evaluations and carbon footprint studies. Versatile studies have proven that geopolymer concrete may be able to decrease greenhouse gas emissions by up to approximately 80% compared to ordinary Portland cement concrete. This success has positioned geopolymer technology within international efforts to mitigate climate change and promote the development of sustainable construction techniques. A considerable amount of research and development has been devoted to geopolymer concrete technology with numerous advances, but many challenges remain. Variations in the mix design and production method are among the most critical issues. Mapping a variety of raw materials and activator solutions across areas complicates the setting of generally applicable standards, limiting their wide application in the building sector. Efforts are being made to develop detailed standards and criteria that would mitigate the quality and performance variations in geopolymer concrete. A key area of focus is to conduct a long-term performance evaluation of geopolymer concrete construction. Although short-term research has indicated that geopolymer concrete is promising, more field trials and long-term monitoring are required to confirm that geopolymer concrete is durable and structurally sound under different environmental conditions. The performance of geopolymer concrete is continually analyzed by researchers owing to the rapid development of testing techniques and prediction models. Despite this past and recent progress, cost-effectiveness remains a key barrier to the broad implementation of geopolymer technology. Manufacturing procedures and activator solutions that are already in use may be more expensive than the traditional production of cement, while using industrial leftovers as source materials has the potential to lower costs, and cement production may be more costly. Current research on optimizing the manufacturing procedures for geopolymer concrete involves alternate activators and economies of scale to make the geopolymer concrete economically competitive. Together with material scientists, civil engineers, and environmental researchers, the subject of geopolymer concrete will become increasingly important as the subject evolves. On the heels of innovation to advance geopolymer-based 3D printing, self-healing geopolymer composites, and nanomaterial applications to increase performance, joint initiatives are forging ahead. This will continue to be the future of geopolymer concrete, ready to solve unmet limits and open new doors in sustainable buildings.

1.4 Significance of GPC

This novel composition and production technique of geopolymer concrete offers significant environmental benefits. Unlike traditional Portland cement, geopolymer concrete does not use high temperature kilns and produces far less CO_2 during the activation process. Therefore, reductions in energy use and greenhouse gas emissions are made possible. This low carbon footprint of geopolymer concrete is in tandem with global efforts to mitigate climate change and diminish the environmental impact of the building sector. Additionally, the inclusion of diverse industrial byproducts within geopolymer concrete alleviates these materials from landfills and reduces raw material demand, thereby saving natural resources. The circular economy strategy encourages waste management techniques for a sustainable process and tackles one of the problems that have persisted with rising industrial waste buildup. In addition, the use of these byproducts affects the unique chemical and physical properties of geopolymer concrete, contributing to the overall performance. The environmental benefits of geopolymer concrete are only a small part of the improved performance of geopolymer concrete. Owing to its superior durability and resistance to chemical assault, it is well suited for use in rough environments, including maritime structures, wastewater treatment facilities, and industrial floors. With its greater resistance to sulfate attack, acid corrosion, and chloride penetration, this can prolong the service life of buildings exposed to harsh environments for extended periods, reducing maintenance costs and the requirement for numerous repairs or replacements. The high abrasion resistance of geopolymer concrete can improve the life of infrastructure that is affected by high traffic. It is very useful for pavements, industrial floors, and other surfaces that are subjected to heavy traffic or abrasive conditions. The potential long-term cost of ownership benefits and reduced resource use in the maintenance and repair of geopolymer concrete structures can be attributed to the increase in endurance. In addition, these concrete products have extremely high fire resistivity, especially for use as fire-retardant construction materials and coatings. Building safety and the potential to save lives are improved by the capacity to withstand high temperatures with little deterioration. This feature is valuable in high-rise buildings, tunnels, and other constructions where fire safety is a real criterion. The mechanical properties of geopolymer concrete, such as compressive strength, tensile strength, and flexural strength, are generally equal to or surpass the mechanical properties of conventional concrete. This allows for more efficient and lighter structures to be constructed, possibly with reduced material use and cost. Another potential benefit of the rapid strength development of geopolymer concrete is that it can help shorten building timeframes, thereby reducing costs and project time. Second, geopolymer concrete exhibits reduced shrinkage and creep compared with ordinary concrete. These characteristics decrease cracking and increase dimensional stability, particularly in large-scale buildings and prefabricated pieces. Enhancing dimensional predictability can improve structural integrity and reduce maintenance in future. The adaptability of geopolymer concrete depends on its aesthetic attributes. However, it may also be developed to yield different colors and textures, allowing architects and designers to have a greater creative license in their designs. The adaptability of geopolymer concrete, as well as its environmental and performance advantages, makes it a desirable material for utilitarian and ornamental use in modern buildings. Research and development are progressing on technology based on geopolymers, and their use is not limited. Progress has been made in the development of the first geopolymer-based repair materials, self-healing geopolymer concrete, and geopolymer-based composites reinforced with different fibers. However, these advancements will enhance the performance of the material and expand its use within the building sector. These enhancements in performance, combined with the potential for long-term cost savings, make geopolymer concrete a practical and attractive option compared to conventional concrete for a range of construction uses. As the construction sector increasingly focuses on sustainability and lasting performance, geopolymer concrete is set to be a significant factor in future development.

1.5 Summary

Geopolymer concrete (GPC) is an advanced and eco-friendly substitute for traditional Portland cement concrete. It is produced through a chemical reaction involving industrial byproducts, such as fly ash or blast furnace slag, along with alkaline activators. GPC offers numerous benefits including reduced carbon dioxide emissions, enhanced mechanical properties, and improved resilience under challenging conditions. Additionally, it exhibits high fire resistance and has the potential to lower material and transportation costs. GPC development began in the late 1970s, and marked progress was made in the 1990s. Current research focuses on fine-tuning mix designs, enhancing curing methods, and establishing standardized testing procedures. The significance of GPC lies in its environmental advantages, remarkable performance, and versatility in construction applications. As the industry seeks to lessen its ecological impact and promote sustainability, GPC is poised to influence the future of eco-friendly construction techniques.

1.6 Highlights

Geopolymer concrete presents a sustainable and innovative alternative to standard Portland cement concrete. It is formulated from industrial waste materials and alkaline activators. Geopolymer concrete significantly reduces the environmental footprint linked to concrete production. It boasts exceptional mechanical properties and resilience to extreme conditions. Geopolymer concrete maintains its structural strength even in high-temperature environments. It is suitable for various applications, including bridges, tunnels, pavements, and marine structures."

References

G.S. Manjunatha, Radhakrishna, K. Venugopal, S.V. Maruthi, Strength characteristics of open air cured geopolymer concrete. Trans. Indian Ceram. Soc. **73**(2), 149-156 (2014)

P. Mehta, Reducing environmental impacts of concrete. Concr. Int. 61–66 (2001)

NRMCA, *Concrete CO$_2$ Fact Sheet* (NRMCA Publication, 2012)

K.K. Poloju, K. Srinivasu, Physical and mechanical properties of calcinated materials under various curing conditions. Math. Stat. Eng. Appl. **71**(3), 1443–1458 (2022a). ISSN: 2094-0343 2326-9865

K.K. Poloju, K. Srinivasu, Influence of GGBS and concentration of sodium hydroxide on the strength behavior of geopolymer mortar. Mater. Today Proc. **65**(Part 2), 702–706 (2022b). ISSN 2214-7853

F. Puertas, S. Martínez-Ramírez, S. Alonso, T. Vázquez, Alkali-activated fly ash/slag cement: strength behavior and hydration products. Cem. Concr. Res. **30**(10), 1625–1632 (2000)

B.V. Rangan, D. Hardjito, S.E. Wallah, D.M. Sumajouw, Properties and applications of fly ash-based concrete. Mater. For. **30**, 170–175 (2006)

S.D. Wang, X.C. Pu, K.L. Scrivener, P.L. Pratt, Alkali-activated slag cement and concrete: a review of properties and problems. Adv. Cem. Res. **7**, 93–102 (1995)

Chapter 2
Elements of Geopolymer Concrete

2.1 General

Geopolymer concrete is a new building material that is environmentally benign and an alternative to conventional Portland cement concrete. The fundamentals of the process: a binder is formed using alkali-activated aluminosilicate ingredients (for instance, fly ash and metakaolin). To activate the aluminosilicate materials, a process known as geopolymerization requires silica and either alumina or silica–alumina species to dissolve in a highly alkaline environment, then condense and rearrange to form a three-dimensional network structure. The geopolymerization process is generally initiated by combining aluminosilicate precursors with alkaline activators like sodium hydroxide (NaOH) or potassium hydroxide (KOH), and frequently sodium or potassium silicate. The aluminosilicate materials break down upon activation in a manner that transports reactive silica and alumina species into the solution. These species are then recombined into oligomers, which finally polymerize into a gel-like structure as their reaction progresses. The final geopolymer matrix is formed when this gel gradually hardens and crystallizes. This binder synthesis process produces a geopolymer binder with superior mechanical qualities, durability, and fire resistance. For strength development, geopolymer concrete relies on the formation of aluminosilicate gel rather than calcium silicate hydrate (C–S–H) gel. Because of its special chemistry and ability to provide high early strength, minimal shrinkage, and—above all—great chemical attack resistance, geopolymer concrete is appropriate for a variety of applications in aggressive environments. The main advantage of geopolymer concrete over traditional Portland cement concrete is that it is much further from generating a significant carbon footprint, making it the perfect choice if you care about it. About 8% of global CO_2 emissions are attributed to the production of Portland cement, mainly because of the calcination of limestone and the high energy requirements for manufacturing. On the other hand, industrial byproducts or nature-based materials are the main binders of geopolymer concrete, reducing the intensive energy use needed to produce clinker and the resulting CO_2 emissions. Such

K. K. Poloju and K. Srinivasu, *Geopolymer Concrete*,
SpringerBriefs in Applied Sciences and Technology,
https://doi.org/10.1007/978-981-96-2479-9_2

a reduction in carbon footprint aligns with global efforts to combat climate change and promote sustainable construction. Geopolymer concrete's mechanical properties can be tailored to meet specific requirements by changing the composition of precursor materials and alkaline activators. The final properties of the geopolymer concrete depend on factors such as the silica-to-alumina ratio, the concentration of the alkaline solution, and the curing conditions. The flexibility of geopolymer mixtures allows for the optimization of suitable mixtures for use in both high-strength structural elements and lightweight insulating materials. Geopolymer concrete is a versatile material that can be fine-tuned to meet a wide range of construction needs. Apart from its excellent mechanical and environmental properties, geopolymer concrete also exhibits good durability characteristics. It possesses superior resistance to acid attack, sulfate attack, and alkali-silica reaction compared to conventional concrete. The thick, amorphous structure of the geopolymer matrix enhances durability by restricting the penetration of aggressive chemicals and inhibiting the formation of expansive reaction products. Geopolymer concrete demonstrates enhanced durability relative to traditional concrete, leading to an extended service life for buildings, less maintenance expenses, and increased sustainability in the constructed environment. Geopolymer concrete exhibits fire resistance, especially where high-temperature stability is essential. In contrast to Portland cement concrete, which experiences a considerable reduction in strength when subjected to heat, geopolymer concrete maintains its structural integrity and elastic modulus up to 800 °C. This attribute renders it advantageous for fire-resistant structures and protective coatings in challenging locations, including fireplaces, skyscrapers, and industrial operations. The extensive benefits of geopolymer concrete are well recognized; nonetheless, obstacles to its broad implementation persist. The unavailability of long-term performance data, particular mixing and curing needs, and the lack of standardized design rules and specifications are significant hurdles. The variability in the composition and reactivity of precursor materials necessitates stringent quality control on both compositional and reactivity standards to provide a uniform end product. These problems need ongoing research and the establishment of industry standards, as well as educational initiatives for construction professionals about the proper utilization and use of geopolymer concrete. Advancements in geopolymer technology are occurring, with the discovery of new applications and enhancements. Recent study has investigated the use of geopolymer in 3D printing, using its quick setting and elevated early strength properties. This discovery has the potential to transform the construction industry by facilitating the fast and efficient fabrication of intricate structural components with minimum waste. Current research focusses on the creation of hybrid geopolymer systems that integrate supplementary cementitious materials, the application of alternative activators to minimize costs and environmental effects, and the examination of geopolymer concrete's efficacy in immobilizing hazardous waste materials. Hybrid geopolymer systems are being created to combine the benefits of geopolymer technology with those of traditional cementitious materials. Research is now being conducted to enhance the performance and reduce the cost of geopolymer concrete by the integration of additional cementitious materials, such as ground granulated blast furnace slag or silica fume. Hybrid systems may compromise some of their rigidity in the transition from concrete to entirely geopolymeric goods, hence promoting swifter acceptance of the technology within the industry. The need to mitigate the environmental and economic repercussions of geopolymer concrete manufacturing has prompted an exploration for alternate activators. Sodium hydroxide and sodium silicate serve as efficient activators; yet their production is costly and energy demanding. Given that geopolymer concrete has a comparatively low environmental impact, researchers are exploring the utilization of more sustainable alternatives, like waste-derived activators or naturally occurring alkaline substances, to further improve the environmental attributes of the concrete. Geopolymer concrete has the capability to immobilize hazardous waste elements, highlighting significant environmental implications for its prospective uses. Research demonstrates that heavy metals, radioactive waste, and other hazardous compounds may be efficiently contained inside geopolymer matrices, hence inhibiting their environmental discharge. These applications may provide a sustainable solution for the long-term storage and disposal of hazardous compounds, a considerable environmental issue. Should the building sector persist in adjusting to the issues presented by climate change and resource depletion, the utilization of geopolymer concrete is expected to rise. As a viable substitute for conventional Portland cement concrete, its distinctive amalgamation of environmental advantages, performance, and superior durability makes it attractive to concrete professionals. Further research and development are essential to actualize the future potential of geopolymer technology, requiring cooperation among academia, industry, and regulatory entities. Geopolymer concrete is a viable sustainable alternative to Portland cement concrete, owing to its distinctive combination of environmental benefits, mechanical properties, and durability. The building sector, in its pursuit of sustainable solutions, may find geopolymer technology crucial in influencing the future of sustainable construction. Subsequent research and development in this domain are expected to provide enhanced performance, cost-efficiency, and user-friendliness, hence facilitating the wider use of geopolymer concrete in diverse building initiatives. Geopolymer concrete, a pivotal technology for advancing a sustainable built environment, may substantially mitigate the environmental effect of the construction sector while meeting the need for durable, high-performance building materials.

2.2 Chemistry of Geopolymer Concrete

Geopolymer concrete has recently garnered attention as an innovative and ecologically friendly building material, perhaps serving as a viable alternative or supplement to standard Portland cement. This sophisticated material employs industrial waste abundant in aluminosilicates, such as fly ash from coal-fired power stations or crushed granulated blast furnace slag from steel manufacture, instead of ordinary Portland cement as the principal binder. The use of waste materials mitigates the environmental effect of concrete manufacturing and overcomes challenges associated with industrial waste disposal.

2.2.1 Chemical Composition of Geopolymers

In a heated setting, the alkaline activation of aluminosilicate molecules produces the geopolymer. Nonetheless, the precise chemical events responsible for the establishment of alkali-activated binders remain incompletely understood. This procedure depends on the particular materials used and the alkaline activator utilized. Davidovits proposed a dual-phase reaction mechanism for geopolymer synthesis: the first chemical reaction of geopolymeric precursors (e.g., aluminosilicate oxides with alkali silicate yielding monomers such orthosilicate ions) followed by the exothermic polycondensation of these monomers. Davidovits asserts that three-dimensional aluminosilicate structures exist in three forms, determined by the silica and alumina concentration of the parent material (geopolymer).

2.2.2 The Fundamental Concept of the Polymerization Process

Raw materials containing aluminosilicates are combined with alkaline activators in the exothermic geopolymerization process, resulting in intricate geopolymers. Rao states that the most often used activators are potassium hydroxide (KOH), sodium hydroxide (NaOH), potassium silicate (Na_2SiO_3), and sodium hydroxide (NaOH). Potassium hydroxide and sodium hydroxide are the only permissible activation agents. The reaction rate is reduced in the presence of only hydroxides. For this reason, Kong (2010) found that hydroxide-silicate combinations (potassium and sodium hydroxide with potassium and sodium silicate) are preferable for accelerating this polymerization process. The dissolution of components in the source material requires a certain concentration of sodium hydroxide. Greater aluminosilicate dissolution results in more polymerization of calcium carbonate. Polymerization begins with the dissolution of solid reactants. During this dissolution, Al_3+ and Si_4+ ions are initially removed from the source materials. A higher leaching rate for aluminosilicate is achieved with sodium-based activators compared to potassium-based ones. Solid particles are better dissolved when sodium silicate is mixed with a high concentration of sodium hydroxide, generating a very alkaline environment, allowing Al and Si atoms to be leached off the solutes, which then undergo reorientation and solidification reactions involving specific structures and contribute to a compacted cemented framework. This occurs because Al–O bonds are weaker and break more readily than Si–O bonds, making reactive aluminum more soluble than silicon. The process continues as the solid particles dissolve more Si–O groups into the solution, increasing the silicate concentration. When the solution contains more silicon, this gel is referred to as Gel 2, a NASH gel (Fig. 2.1).

Geopolymers are reaction-activated materials that need energy input, such as thermal curing in an oven. This investigation included the preparation of geopolymer mortar in furnaces and its curing at ambient temperature. The characteristics of

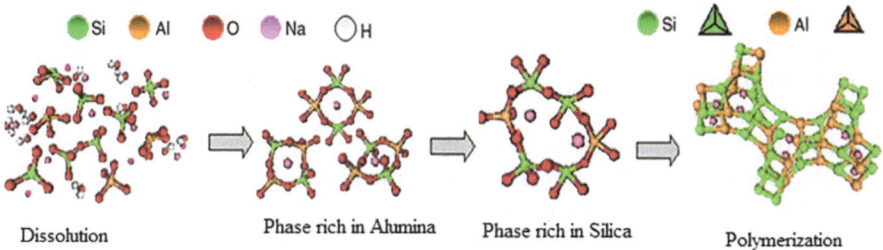

Fig. 2.1 Schematic diagram of the polymerization process

geopolymers are affected by the molar ratio of Na_2SiO_3 to NaOH. Test results indicate that the maximum compressive strength calculated for constant binder content is 2.5%. It was also found that heating GPC containing fly ash at room temperature increased strength over time, significantly more than GPC without fly ash. The impact of modifying the Na_2SiO_3 to NaOH ratio (1:1) while sustaining a sodium hydroxide concentration of 10 M was examined, revealing that compressive strength enhanced with the inclusion of GGBS and alkaline content. Research indicates that geopolymer serves as a better binder for concrete owing to its enhanced strength, durability, cost-effectiveness, and environmental sustainability.

The behavior of the mortar's strength is being analyzed. Any other alkaline solutions besides NaOH and Na_2SiO_3 were deemed inadequate for producing geopolymer concrete. Silicon and aluminum combined in fly ash favor faster dissolution and increased compressive strength. GGBS can be activated by an alkaline solution. Consequently, GGBS is used as a replacement for fly ash owing to its capacity to expedite polymerization and improve the strength characteristics of concrete. The chemistry of geopolymer concrete is intricate and fascinating, including the reactivity of aluminosilicate minerals with alkaline solutions of sodium hydroxide or potassium hydroxide, often combined with sodium silicate. The selection of alkaline activator and its concentration are critical factors influencing the ultimate characteristics of geopolymer concrete. The geopolymerization process is initiated by the alkaline activator. This interaction results in a sequence of dissolution, reorientation, and polycondensation events that ultimately produce a three-dimensional aluminosilicate network structure. Upon dissolution of the aluminosilicate, it disintegrates in the alkaline solution, releasing silicon and aluminum ions into the mixture. This technique requires the fast decomposition of aluminosilicate compounds, attainable just in an alkaline environment produced by the alkaline activator. Upon reorientation, these ions commence polymerization, resulting in the formation of short molecular chains (polymers). These oligomers undergo polycondensation, resulting in a resilient, interconnected network of polymerized aluminosilicate polymers that facilitates the process. The procedure is contingent upon variables like temperature, curing conditions, and the silicon-to-aluminum ratio in the source materials. The resultant geopolymer binder demonstrates superior mechanical capabilities for compressive, tensile, and flexural strength. The mix design, curing conditions, and the kind and concentration of alkaline activators may be modified to optimize these qualities to satisfy the specified criteria. High early strength is a defining quality of geopolymer concrete, which may attain superior early strength compared to ordinary concrete, usually during the first 24–48 h post-casting. In precast concrete applications and other scenarios requiring expedited building timelines, the quick development of strength is advantageous. The initial strength development is ascribed to the fast establishment of the aluminosilicate network structure during the geopolymerization process. In addition to its remarkable mechanical capabilities, geopolymer concrete has superior durability relative to traditional concrete. The material's thick microstructure and distinctive chemical composition are thought to enhance durability by providing exceptional protection against several types of deterioration. A common weakness seen in traditional concrete, namely the calcium hydroxide hydration product from Portland cement, is not present in geopolymer concrete. Geopolymer concrete has exceptional resistance to acid attack and is well suited for hostile situations, including industrial buildings and wastewater treatment plants. All of the aforementioned indicates that geopolymers exhibit a more stable aluminosilicate network structure under acidic conditions than the calcium-rich phases seen in conventional concrete. This also signifies enhanced resistance to sulfate assault, which is essential in maritime situations or regions with sulfate-rich soils. The absence of sulfate degradation in this cement is attributable to the lack of tricalcium aluminate, a constituent of Portland cement vulnerable to sulfate assault. Moreover, it has been shown that geopolymer concrete has much superior resistance to alkali-silica reaction (ASR), a harmful expansion reaction in concrete resulting from the interaction between alkaline cement and certain particles. Geopolymer concrete exhibits resistance to alkali-silica reaction (ASR) because of its reduced calcium content and distinct pore solution chemistry. Geopolymer concrete demonstrates resilience to elevated temperatures. Traditional concrete deteriorates at temperatures beyond 300 °C, but geopolymer concrete maintains its structural integrity at temperatures reaching 800 °C. It is an ideal material for applications including fire-resistant constructions, refractory materials, and environments characterized by elevated temperatures in industrial contexts. The enhanced heat resilience of geopolymer concrete is due to its ceramic-like structure and the lack of hydration products, which breakdown at temperatures above the combustion threshold of traditional concrete. Moreover, geopolymer concrete has ecological advantages. The use of industrial wastes as the principal binder enables geopolymer concrete to diminish the carbon impact of building operations. Portland cement, a fundamental component of conventional concrete, constitutes around 8% of total world CO_2 emissions. The energy-demanding procedure of clinker manufacturing and the calcination of limestone substantially contribute to this environmental effect. The carbon emissions linked to the production of geopolymer concrete may be as much as 80% lower than those of traditional Portland cement concrete, contingent upon the mix design and raw ingredients used. The elimination of clinker manufacture, which is often fuel-intensive and generates significant CO_2 emissions, facilitates the use of industrial waste materials, thereby decreasing carbon emissions. Furthermore, the advantageous use of substantial quantities of industrial wastes, such as fly ash and slag, often stored or disposed of in landfills, presents prospects for geopolymer concrete. This not only mitigates the environmental impact of waste disposal but also preserves natural resources in concrete manufacturing by lowering the need for virgin raw materials. Moreover, it has been shown that geopolymer concrete may be used for the immobilization of heavy metals and other toxic chemicals in certain industrial wastes, hence aiding in environmental preservation. Notwithstanding the many benefits of geopolymer concrete, its extensive implementation encounters obstacles. This encompasses intricate mix design optimization, the possibility of elevated material expenses in certain areas, and the absence of long-term performance data relative to conventional concrete. In contrast to conventional concrete, the mix design of geopolymer concrete requires consideration of additional elements, such as the kind and concentration of the alkaline activator, the chemical composition of the aluminosilicate source materials, and the curing conditions. If inadequately managed, this complexity may result in performance fluctuation. The accessibility of raw ingredients, especially alkaline activators, might vary considerably across various places owing to geographic variables, thus diminishing the economic feasibility of geopolymer concrete in some regions. Furthermore, the elevated alkalinity of alkaline activators necessitates specific safety measures for their handling and storage in certain building environments. A further impediment to the adoption of important infrastructure projects, where enduring durability is paramount, is the absence of long-term performance data and set criteria for geopolymer concrete. Nonetheless, initiatives are underway to tackle these challenges, with current research and development concentrating on the use of geopolymer concrete in diverse building endeavors.

Researchers are investigating other sources of raw materials outside natural pozzolans, including industrial leftovers, to broaden the raw material basis for geopolymer manufacturing. Initiatives are underway to develop more user-friendly alkaline activators and to refine mix formulations for enhanced workability and uniformity. Extended field investigations and demonstration projects of geopolymer concrete are now being conducted under diverse environmental circumstances. Standards and requirements for geopolymer concrete are currently being developed, as nations and organizations strive to create criteria for its use in building. These initiatives are essential for instilling trust in engineers, architects, and regulatory bodies on the viability of geopolymer concrete as an alternative to traditional concrete. The popularity of geopolymer concrete is expected to increase as further evidence emerges and standards are established. In summary, geopolymer concrete is a viable substitute for conventional Portland cement-based concrete, demonstrating superior mechanical, durability, and ecological characteristics. Owing to their distinctive chemical composition and microstructure, these materials exhibit exceptional resilience to many types of deterioration, making them particularly suitable for demanding settings where traditional concrete may falter. This material's high early strength and maintained structural integrity at increased temperatures enable its use in diverse building projects. Geopolymer concrete has environmental benefits, notably its capacity to substantially decrease CO_2 emissions during concrete manufacturing, establishing it as a viable technology for sustainable building methodologies. Geopolymer concrete is poised to assume a more prominent position in the evolution of building materials, as the construction sector pursues sustainable alternatives to address escalating global infrastructure requirements while reducing environmental effect. Nonetheless, enhancements are still required in mix design optimization, material availability, and long-term performance data, underscoring the need of continuous efforts to address these challenges. As our comprehension of geopolymer chemistry and its long-term performance advances, and as standards and requirements are established, the utilization of geopolymer concrete is anticipated to increase significantly. This novel material has the potential to transform the building industry by offering a more sustainable and lasting alternative to standard concrete, while also supporting the circular economy via the advantageous use of industrial leftovers. Geopolymer concrete has significant promise, transcending traditional construction applications to include 3D printing of edifices and sophisticated composite materials. Geopolymer concrete is poised to become a crucial component of a sustainable future in building as research advances and practical experience evolves. Its mechanical qualities, improved durability, and ecological advantages make it a formidable contender for addressing the issues of contemporary building. The distinctive chemical composition and microstructure provide exceptional resilience to several kinds of deterioration, making it appropriate for use in demanding settings where traditional concrete may falter. The capacity to attain significant early strength and preserve structural integrity at high temperatures creates new opportunities for its use in specialized building endeavors. The environmental advantages of geopolymer concrete, particularly its ability to significantly reduce CO_2 emissions associated with concrete production, make it an essential technical innovation in the pursuit of sustainable construction practices. Geopolymer concrete is poised to significantly influence the future of construction materials as industry pursues sustainable alternatives to address the growing need for global infrastructure. Despite ongoing obstacles concerning material availability, mix design optimization, and long-term performance data, these concerns are progressively being addressed via sustained research and development efforts. The utilization of geopolymer concrete is anticipated to rise as standards and specifications are established and as our comprehension of geopolymer chemistry and its long-term performance improves. This innovative material has the capacity to revolutionize the building industry by offering a more durable and sustainable substitute for conventional concrete, while fostering the circular economy via the beneficial use of industrial waste. Geopolymer concrete exhibits promising prospects, with potential applications transcending traditional construction to encompass advanced composite materials, 3D printing of edifices, and extraterrestrial endeavors, such as lunar or Martian exploration, where utilizing local regolith as a raw material may prove essential. Geopolymer concrete is set to significantly contribute to sustainable building in future, if research and practical applications progress.

2.3 Variations of Geopolymer Concrete

An innovative and sustainable substitute for conventional Portland cement concrete is geopolymers. This binder, which is made by alkaline activation of aluminosilicate minerals, has exceptional mechanical qualities and durability. Recently, this sustainable construction material has caught the heat owing to its potential to cut carbon emissions that are rampant in the construction industry.. The manufacture of geopolymer concrete, as a viable approach for mitigating climate change issues, emits up to 80% less CO_2 compared to traditional concrete. Geopolymer concrete is categorized into many categories according to their source ingredients and activation techniques.

Geopolymer concrete derived from fly ash:

- Primarily use fly ash as the principal raw material, a byproduct of coal combustion.
- High compressive strength and acid attack resistance.
- Must be heat-cured to achieve optimum strength development.
- Has low shrinkage and creep.
- It is an effective way to use industrial waste and thus lower the need to end up in landfill.
- The rigidity can be designed to be up to 100 MPa or more in strength.
- Its excellent resistance to sulfate attack and alkali-silica reaction has been demonstrated.

Geopolymer concrete based on ground granulated blast furnace slag (GGBS):

- GGBS, a byproduct of steel production, is the primary precursor.
- It offers excellent workability and time-set control.
- It provides high early strength and improved durability.
- It can be cured at ambient temperatures.
- It exhibits lower heat of hydration than Portland cement concrete.
- It shows enhanced chloride permeability resistance and is therefore suited for marine environments.
- Provides better freeze–thaw resistance in cold climate sites.

Metakaolin-based geopolymer concrete:

- It uses metakaolin, a calcined form of kaolin clay, as the main aluminosilicate source.
- High mechanical strength as well as fire resistance.
- Low shrinkage and good resistance to aggressive environments.
- Workability needs to be controlled, requiring careful mix design.
- Dimensional stability is good, and efflorescence is reduced.
- Exhibits superior resistance to acid attack, relative to other geopolymer types.
- Used for high-performance applications with improved durability.

Red mud-based geopolymer concrete:

- It uses red mud, a byproduct from bauxite processing, as a partial precursor.
- It is important in the sustainable management of industrial waste.
- It shows good mechanical properties and durability.
- Optimizing mix design is required to address potential problems with setting time and workability.
- Red mud disposal is reduced by its contribution to minimizing environmental impact.
- Further treatment or amalgamation with other materials may be necessary to alleviate any pollutants.

Geopolymer concrete including rice husk ash:

- This is an alternative cementitious material that utilizes rice husk ash, an agricultural byproduct.
- Enhances workability and lowers water requirements.
- It improves resistance to chloride penetration and sulfate attack.
- It helps reduce agricultural waste.
- It is a highly reactive silica source for the geopolymerization process.
- Excellent pozzolanic properties are shown, enhancing long-term strength development.
- It can be used to produce high-performance durable concrete.

Hybrid geopolymer concrete:

- Optimizes properties by combining two or more source materials (fly ash and GGBS).
- It tailors mechanical and durability characteristics.
- Provides flexibility in raw material selection concerning availability.
- This potential reduces the requirement for heat curing.
- It permits the development of geopolymer concrete with enhanced performance in certain applications.
- It provides opportunities for synergistic effects between different precursor materials.
- It enables the optimization of mix design to develop required fresh and hardened properties.

Natural pozzolan geopolymer concrete activated by alkali:

- The primary source of aluminosilicate in alkali-activated natural pozzolan geopolymer concrete is natural pozzolans, usually pumice or volcanic ash.
- It provides good mechanical properties and durability.
- An alternative is regions with limited industrial by-products that are sustainable.
- The optimal composition of the activator can be optimized for best performance.
- Has excellent resistance to aggressive environments, i.e., acid and sulfate attack.
- It can be used to make high-strength concrete under good mix design and curing conditions.
- The possibility for reduced environmental impact compared to Portland cement concrete.

Geopolymer foam concrete:

- Arrives in foam or with lightweight aggregates to produce a low-density material.
- It has excellent thermal insulation properties.
- It is a good fire-resistant and sound-absorbing material.
- For non-structural applications and precast applications.
- Potential for use in lightweight panel and block production for construction.
- It shows promise for lowering energy use in buildings because of improved insulation.
- It provides options to create energy-efficient and sustainable construction materials.

All types of geopolymer concrete have specific properties, advantages, and disadvantages. The best type depends on local availability of raw materials, quality properties, climatic conditions, and economics. Several combinations of precursor materials and activation methods are ongoing studies to boost the properties of geopolymer concrete as well as sustainable improvement.

1. **Reducing impact on the environment**: With millions of cubic meters produced all over the world, geopolymer concrete technology has developed, and numerous construction projects have successfully employed it. It spans paving, bridge work, precast components, and even watertight buildings. When mature and with wider penetration on the construction site, this needs to be an important part of reducing the impact on the environment and also achieving a sustainable built environment.
2. **Overlaying the potential long-term performance of geopolymer concrete**: The challenges still exist to its widespread recognition, and these arguments include standardization, data on long-term performance, and education of industry professionals. At present, by investigating the production processes of geopolymer concrete and solutions for obtaining better properties, it has become highly necessary to overcome these challenges in order to utilize them significantly for construction applications.
3. **Alkali-silica reaction (ASR)**: The reduced calcium content and distinct pore solution chemistry in geopolymer concrete mitigate the likelihood of ASR. The improved resilience may result in an extended service life and decreased maintenance expenses for buildings constructed using geopolymer concrete.
4. **Fire resistance**: When it comes to fire resistance, geopolymer concrete is more advanced compared with typical Portland cement concrete. It can withstand temperatures up to 800 °C without compromising stability. This behavior results from the ceramic characteristics of the geopolymer binder, which, rather than disintegrating like calcium-based cement paste, undergoes sintering at elevated temperatures. The superior fire resistance of geopolymer concrete makes it a desirable construction material for situations where fire safety is paramount, including high-rise structures and tunnels.
5. **Heat properties**: Geopolymer concrete has lower heat conductivity than traditional concrete, perhaps leading to enhanced insulating capabilities. This will

contribute to better energy efficiency of buildings. This variation in pore distribution and hyper glyceride, as compared to conventional cement concrete, might cause its lower thermal conductivity. However, since conventional concrete has a higher coefficient of thermal expansion compared to this material, if used for making large members, they will induce lower thermal stresses.

6. **Shrinkage and creep**: Compared to standard concrete, geopolymer concrete tends to show less drying shrinkage as well as reduced creep. This trait can make it less cracked and steadier in the long run. This relatively low shrinkage, compared to more traditional concrete, is due in part to the lack of calcium silicate hydrate gel that accounts for a significant portion of ordinary concrete shrinkage. The less creepy behavior is associated with the stiff three-dimensional network structure of the geopolymer binder. As a result, these characteristics render geopolymer concrete highly appropriate for any kind of work requiring dimensional stability, such as precast concrete components and mega-infrastructural projects.

7. **Working time**: The setting time of geopolymer concrete may be adjusted by mix design and activator concentration. It may be engineered for either a rapid or prolonged setting time, depending on the intended use. The required setting time may vary depending on the building context, ranging from rapid repair tasks to extensive casting processes; this adaptability renders it advantageous. The setting of geopolymer concrete primarily results from the dissolution of aluminosilicate species, followed by polycondensation, a process influenced by factors such as the kind and quantity of alkaline activators, the reactivity of source materials, and ambient temperature.

8. **Workability**: it might be affected by the type and amount of alkaline activators. The workability is mainly dependent on the viscosity of the alkaline solution as well as the particle size distribution of source materials. Geopolymer systems can also be used for inline high-performance superplasticizers that improve workability without decremental or variable strengths. It is essential to recognize that geopolymer concrete often has a diminished open time compared to conventional concrete, necessitating meticulous planning for installation and finishing procedures.

9. **Bond strength**: Geopolymer concrete often exhibits strong bond strength with reinforcing steel, a crucial characteristic for its use in reinforced concrete. This bond strength is affected by the chemical composition of the geopolymer binder and by interfacial properties between the binder and reinforcement. Geopolymer concrete has been shown in some studies to have better bond strength than when Portland cement concrete is used, especially at high temperatures. This characteristic inhibits any slippage between concrete and reinforcement by transmitting stresses via the adhesive layer at their interface, which is crucial for the overall efficacy of reinforced geopolymer concrete components.

10. **Environmental advantage**: The primary environmental advantage of geopolymer concrete is its reduced carbon emissions in comparison to the Portland cement concrete process. The absence of cement clinker production can reduce CO_2 emissions by up to 80%. The reengineering of geo-materials enabled

by this approach leads to reduced carbon emissions, mainly due to the use of industrial side-streams or waste as precursors and the lower energy demand to produce activator materials compared to cement. Geopolymer concrete is an innovative construction material that helps mitigate the harmful impacts of the building industry on carbon emissions around the world and facilitates a transition to green building.

11. **Resistance**: Geopolymer concrete offers excellent resistance to a range of chemicals (moreover, even against acidic environments), enabling its use for implementations in aggressive environments such as wastewater treatment plants and many industrial facilities. Organisms responsible for biocorrosion of concrete can damage the interfacial regions of traditional concrete and increase permeability to liquid ingress over time; chemical stability, on the other hand, is related to work by others demonstrating that geopolymers have a more stable three-dimensional aluminosilicate network structure compared with calcium-rich phases of blended binders. The enhanced chemical resistance is also due to geopolymers compared with those traditional cement products. It has long been known that the faster degradation rate of cement products is less resistant to acid attack or other chemicals. Due to this property, geopolymer concrete could be an attractive performance-based option for structures subjected to aggressive chemical conditions in terms of improving service life and lowering maintenance requirements.

12. **Permeability**: Geopolymer concrete generally exhibits reduced permeability compared to traditional concrete, hence improving its longevity and resistance to chemical degradation. The limited permeability results from the dense microstructure established during the geopolymerization process. This reduced permeability limits the ingress of detrimental substances such as chlorides and sulfates, hence enhancing the longevity of concrete in practical applications. This attribute is particularly advantageous in maritime settings or in structures exposed to de-icing salts.

13. **Flexural strength**: Geopolymer concrete often has commendable flexural strength, frequently comparable to or exceeding that of conventional concrete. Factors affecting flexural strength include the kind and quantity of aluminosilicate components, the composition of alkaline activators, and the circumstances of curing. Geopolymer concrete, demonstrating excellent flexural properties, is suitable for applications subjected to significant bending forces, including beams, slabs, and pavements.

14. **Modulus of Elasticity**: The elastic modulus of geopolymer concrete may vary (be lower or comparable) to that of conventional concrete, contingent upon the mix design classifications. The elastic modulus is a critical property for structural design since it governs deformation behavior. The somewhat reduced elastic modulus of geopolymer concrete may be beneficial in certain applications owing to improved stress distribution and fracture resistance.

15. **Curing**: Conventional concrete attains strength by ambient heat, but geopolymer concrete requires increased temperature curing for maximum compressive strength growth; nonetheless, research is focused on creating

ambient curing techniques for geopolymer mixes. One of the major issues with the widespread use of geopolymer concrete has been the requirement for heat curing, especially when used to make cast-in-place applications. Most of the recent literature is concentrated on providing geopolymer mixes that satisfy enough strength under ambient curing conditions. This innovation is essential for broadening the application of geopolymer concrete in different construction situations.

Finally, geopolymer concrete has a series of properties that provide it with considerable advantages over conventional concrete in several applications. Not to mention, it is much more durable, fireproof, and environmentally friendly. With further development in this field tackling difficulties like curing at ambient temperature and long-term performance data, geopolymer concrete will have a wider application range, helping to improve our construction methods.

Geopolymer concrete has outstanding characteristics that lead to new structural designs and construction practices. The early strength and low shrinkage nature of these cements are advantageous in precast concrete production, where rapid production cycles and dimensional stability are crucial. The good fire resistance of geopolymer concrete can change the design of high-rise buildings and tunnels with the possibility of eliminating extra fire protection.

Furthermore, the characteristics of geopolymer concrete embody the ideas of a circular economy within the building industry. Geopolymer concrete is an innovative eco-friendly building material that employs industrial by-products such as fly ash and crushed granulated blast furnace slag, therefore preserving virgin raw resources and creating a market for things often regarded as trash. The nature of geopolymer concrete production aids toward resource efficiency and large-scale waste minimization. With growing pressure on the construction industry to address environmental challenges, geopolymer concrete offers a promising option for sustainable infrastructure development. It also has a potential major role in reducing the contribution of large-scale production to carbon emissions that will need to be achieved if climate targets are to be met globally. Nevertheless, for universal acceptance of geopolymer concrete as a structural material, more research is needed, and standards/specifications must be developed together with exercises in educating engineers and construction professionals regarding the properties of this new material and its application.

Geopolymer concrete technology is a developing field, so future directions include:

- The creation of high-strength geopolymer mixes by carefully balancing the composition throughout and achieving ambient curing conditions.
- Developing geopolymer concrete for use in different environmental conditions: a review of the long-term performance of geopolymer concrete structures.
- Mix designs customized to end applications (for example, concrete 3D printing, ultra-high-performance concrete).

- Investigating new sources of aluminosilicate materials to broaden the geopolymer matrix raw material base.
- Creating low-cost and eco-friendly alkaline activators to make the environmental footprint of geopolymer concrete even lower.
- With the continuing research in these aspects, geopolymer concrete can emerge as one of the predominant materials for construction that is not only sustainable but also a high-performance substitute for Portland cement concrete. Informational and geopolymer concrete, with its unique properties and environmental benefits, can be considered one of the important technologies for the shift of our infrastructure projects toward larger sustainability and resilience goals.

The potential for erosive patterns, specific real-world examples, and case studies, such as from a macroeconomic perspective, the adoption of geopolymer concrete could have important economic implications for construction practices. The lower carbon footprint of geopolymer concrete could be a competitive advantage where environmental regulations become tighter, and soon carbon pricing schemes are implemented in many regions. It might even alter the way concrete is processed and result in new jobs upstream or deliver innovative solutions downstream.

Moreover, geopolymer concrete may enhance infrastructure. It is more resilient and resistant to environmental degradation, implying that structures may endure longer with reduced maintenance, thereby saving substantial costs throughout the lifecycle of infrastructure projects and enhancing the overall sustainability of built environments.

Geopolymer concrete is an alternative with several uses outside traditional buildings. Its distinctive features render it suitable for certain applications, such as nuclear waste encapsulation, attributable to its excellent chemical resistance and low permeability.

3D printing of concrete buildings, with adjustable setting time and excellent layer adhesion capabilities.

- Utilization of its high rate of strength increase and excellent bond ability to existing concrete for the repair and rehabilitation of in-situ structures.
- Marine and offshore structures, having high resistance to chloride ingress and sulfate attack.
- Low-density high-temperature applications, using it for its excellent fire resistance property (furnace linings or chimney flues).

With ongoing research into geopolymer technology, more uses for this material are sure to be found, broadening its potential influence on the construction industry and other fields even further. On the other hand, some challenges also exist to promote the widespread use of geopolymer concrete. These include:

- Quality control and standardization of raw material source variety.
- Alkaline activators are not yet competitive with traditional cement in terms of price, though this may change as production picks up.
- Limited long-term environmental performance data of geopolymer concrete structures.

- Specialized apparatus and training for management, together with the caustic properties of alkaline activators.
- The potential for property variances contingent upon mix design and curing circumstances, which must be meticulously regulated and observed.
- Addressing these difficulties will need collaboration among scholars, industry professionals, and regulators. Finally, proper standards and specifications for geopolymer concrete need to be developed to create a degree of confidence among engineers and clients regarding its employment in construction.
- Geopolymer concrete is a suitable and major technological upgrade in construction materials. This blend of sustainability with world-class performance makes it one of the leading materials in future-proofing construction. Geopolymer concrete could reshape the construction sector through further studies and experience in producing geopolymer concrete, which can become a major tool for stronger, sustainable, and more resilient infrastructure across the globe.

2.4 Summary

Geopolymer concrete has several benefits over conventional cement, including enhanced durability, superior fire resistance, and less environmental impact. The distinctive characteristics of concrete, including its elevated compressive strength, durability, and low permeability, render it suitable for several uses. Yet barriers such as elevated temperature curing and limited long-term performance data must be overcome first before widespread use can take place. The exploration continues with ambient curing, mix design optimization, as well as new applications. These properties of geopolymer concrete can play a significant role in sustainable construction and making resilient infrastructure.

2.5 Highlights

- Lower shrinkage and creep of geopolymer concrete.
- The setting time of geopolymer concrete can be controlled by proper adjustment in the mix design or concentration of the activator (10).
- Geopolymer concrete has good bond strength with steel, which is important in reinforced concrete structures.
- Geopolymer concrete contains up to 80% less CO_2 emissions than Portland cement concrete.
- With extreme resistance to low pH, such as acid attack, geopolymer concrete is fit for aggressive environments against chemical action.
- The permeability of geopolymer concrete is lower than ordinary cement, which helps to enhance its durability.

- Thus, geopolymer concrete is known to have good flexural strength, often at par or higher compared with conventional concrete.

Reference

D.L. Kong, J.G. Sanjayan, Effect of elevated temperatures on geopolymer paste, mortar, and concrete. Cem. Concr. Res. **40**(2), 334-339 (2010)

Chapter 3
Potential Uses of Geopolymer Concrete

3.1 General

Benefits: Geopolymer concrete has many benefits compared to conventional concrete. GPC has the potential to become an alternative infrastructure material because it has several advantages, such as very high resistance against acid (sulfate) attack, good early strength gain, moderate drying shrinkage, and low creep. Geopolymer concrete is recognized as a preferable alternative to traditional concrete (Song et al. 2005; Duxson et al. 2007a, b). Geopolymer represents the effective use of various waste materials. Consequently, the most compelling option for the concrete sector to meet the present CO_2 targets is to transition to sustainable materials.

Although geopolymer concrete has many benefits, as mentioned previously, it has critical drawbacks that should be overcome for its large-scale implementation. The most serious problem is how GPC is managed: for its activation, expensive and harmful chemicals are needed, which must be processed in the right way. The second problem is the sensitivity of the polymerization process. Studies by many researchers on factors affecting the strength and workability of GPC have yielded conflicting findings. Therefore, further research in geopolymer concrete is needed. Unless such a study yields consistent data on the properties of GPC and its production, replacing OPC—which has circulated worldwide since its genesis in the 1840s—will be impossible. Though there are more downsides to using GPC than upsides, it is a commonly used application.

3.2 Applications of Geopolymer Concrete

GPC progresses due to its demand for high temperature curing conditions; hence, it is best suited for precast applications. Fly ash-based geopolymer concrete (GPC) is utilized for manufacturing railway sleepers, sewage mines, structural members,

© The Author(s), under exclusive license to Springer Nature Singapore Pte Ltd. 2025   27
K. K. Poloju and K. Srinivasu, *Geopolymer Concrete*,
SpringerBriefs in Applied Sciences and Technology,
https://doi.org/10.1007/978-981-96-2479-9_3

and retrofitting, as it bonds well with common cement concrete. Composites have been used to strengthen reinforced concrete elements such as beams. Furthermore, it can be used in emergency reconstruction and restoration of damaged constructions using GPC. GPC has been used in Australia for the manufacture of box culverts, bridge decks, railway sleepers, wall panels, retaining walls, and water tanks. The GPC training began with the "Global Change Institute Building" at the University of Queensland in Australia. Geopolymer foam concrete has also been claimed to be useful for thermal insulation.

Geopolymer concrete applications cover spheres from recycling to construction, proving its potential as a fundamental part of an industry-wide change.

The applications of GPC in Civil Engineering are not limited to:

1. **Infrastructure**: Geopolymer concrete, with enhanced durability and environmental resistance, may improve bridges, highways, and tunnels. An improved strength-to-weight ratio enables it to extend greater distances than steel while necessitating less maintenance. Its resilience to extreme circumstances like freeze-thaw cycles and chemical exposure renders it highly favored for critical infrastructure, even in the most severe regions. Moreover, the low permeability of geopolymer concrete reduces water infiltration, hence decreasing the likelihood of reinforcing corrosion, which enhances durability and extends the service life of buildings.

2. **Marine structures**: Geopolymer concrete enhances the durability of structures in harsh marine environments, such as seawalls, offshore platforms, and port facilities, owing to its exceptional resistance to chloride intrusion and sulfate assault. This kind of resilience is crucial in coastal areas, where seawater may severely damage traditional concrete and lead to its continuous deterioration. The enhanced resilience of geopolymer concrete in these conditions might diminish maintenance expenses and extend the lifespan of maritime constructions. Additionally, its ability to gain early strength will help accelerate construction in tidal zones, reducing the exposure of fresh concrete to seawater.

3. **High-rise buildings**: The compressive strength and creep reductions of geopolymer concrete in foundations, columns, and load-bearing walls allow for taller, more economical building designs. The rapid early strength gain of the material allows construction schedules to go faster, resulting in shorter project durations and less associated costs. Geopolymer concrete has low shrinkage properties, which make it best suited to achieve rigidity and strength for tall structures with increased dimensional stability. The fire-resistant characteristics of the material further contribute to the safety of high-rise buildings by providing enough time for evacuation during fire incidents.

4. **Precast elements (beams, panels, pipes)**: The low drying shrinkage and superior dimensional stability of geopolymer concrete can also result in better quality precast products for less energy. This exceptional early strength helps in shorter production cycles at a precast plant, resulting in increased productivity. The flexibility of the geopolymer concrete mix design also allows lightweight precast members with better thermal insulation. Its durability against harsh environments

makes it ideal for precast shapes used in sewage systems, chemical storage tanks, and other challenging applications.

3.3   GPC Utilization in Construction

1. **Eco-friendly construction materials**: Geopolymer concrete serves as an eco-friendly substitute for conventional concrete, potentially reducing carbon emissions from cement manufacturing by up to 80% by using industrial by-products as its basic ingredients. Geopolymer concrete may enhance a building project's sustainability ratings and ensure compliance with increasingly stringent environmental standards. The production procedure allows for the use of recycled resin, enhancing its environmental sustainability. The reduced carbon footprint of geopolymer concrete corresponds with the global initiative for more stringent reductions and fosters a circular economy in building.

2. **Structures resistant to fire**: The improved thermal stability of geopolymer concrete contributes to building safety in residential and other structures by ensuring structural integrity and preventing collapse during fire incidents. This characteristic is particularly crucial in high-risk environments, such as industrial facilities, storage sites, and skyscrapers. The materials' endurance to intense temperatures without significant breakdown may provide valuable additional time for evacuation and fire suppression. The fire resistance of geopolymer concrete leads to reduced insurance costs for buildings and improves safety standards in the construction industry.

3. **Rapid construction**: Owing to its fast-setting time and early strength development, geopolymer concrete may accelerate casting operations, facilitating project completion in a reduced timeframe, hence yielding labor cost savings. It is especially advantageous for time-sensitive projects, such as emergency repairs or infrastructure enhancements in metropolitan areas, when minimizing interruption is crucial. Moreover, it facilitates the expedited removal of formwork owing to its rapid strength development, hence accelerating the project timeline. Geopolymer concrete has significant strength at ambient temperatures, eliminating the need for heat curing and facilitating onsite casting of buildings across many climates.

4. **Architectural components and facades**: Geopolymer concrete may be molded into almost any form, augmenting a building's aesthetic appeal and incorporating many colors that improve its visual attractiveness. The material's adaptability facilitates a diverse array of textures and treatments, enabling architects and designers to innovate in both interior and exterior applications. Geopolymer concrete may be engineered to replicate genuine stone or other materials, providing a cost-effective alternative for premium architectural design. Its capacity to withstand UV radiation and environmental degradation guarantees that the décor retains its hue over time and requires minimum upkeep or replacement.

3.4   GPC for Environmental Remediation Purposes

1. **Waste encapsulation**: Due to its unique chemical properties, geopolymer concrete may effectively connect with hazardous and radioactive waste, securely containing it as a safer long-term alternative to just burying contaminated materials. Its low permeability and excellent chemical resistance provide it an appropriate matrix for the efficient encapsulation of harmful compounds, therefore mitigating environmental contamination. It is particularly relevant in the nuclear sector and the remediation of industrial waste sites. The increased chemical stability of the linkages in geopolymers created with various other wastes results in a decreased leaching potential.

2. **Soil stabilization**: Geopolymer-based binders provide solutions for the treatment of polluted soils and sediments in brownfield remediation and contaminated land management.

3. **Effectiveness in heavy metal immobilization**—The geopolymerization process has been found to effectively stabilize heavy metals and other contaminants through potential binding, thereby reducing mobility and the potential for leaching to groundwater. It is a low-cost and eco-friendly method of site cleanup, allowing for development in previously non-developable spaces. Soil stabilization using geopolymer also increases the strength of treated soils, making them more resistant to loads and erosion, etc.

4. **Carbon sequestration**: Geopolymer concrete has also been found to have the ability to capture and store carbon dioxide, which could present a potential way to mitigate greenhouse gas emissions from the construction sector. Numerous investigations are underway to develop geopolymer compositions capable of naturally sequestering carbon dioxide (CO_2) from the atmosphere over their operational lifespan, so transforming buildings and infrastructure into CO_2 sinks. This characteristic may significantly contribute to climate mitigation and carbon neutrality objectives in the built environment.

5. **Introduction to carbon-negative GPC climate-digital Amsterdam forum**: Geopolymer concrete (GPC) formulations that sequester more CO_2 than they emit in their life cycle could, if developed and adopted at larger scales through industry practice, transform the construction materials sector with respect to sustainability and significantly contribute to combating climate change on a global scale.

6. **Geopolymer components for acid-resistant structures**: Wastewater treatment plants and their industrial facilities can take advantage of the better acid resistance of geopolymer concrete, resulting in an increased lifespan of these infrastructures with hefty investments against corrosion. This level of resistance is especially useful in chemical processing, food production, and mining applications where exposure to acidic materials tends to be high. This may substantially decrease maintenance expenses and downtime resulting from the degradation of concrete forms exposed to acid attack, if geopolymer concrete may be used. Furthermore,

its structural integrity in acidic conditions enhances worker safety and mitigates the danger of environmental contamination resulting from structural failures.

3.5 Supplementary Applications

1. **Aerospace**: Due to the exposure of several spacecraft and aircraft components to severe temperatures and thermal cycling, geopolymer concrete may serve as a viable option for aerospace sector components to enhance performance under such circumstances. Its low thermal conductivity and high thermal stability, along with its temperature resistance and corrosion resistance, make the material suitable for use in heat shields, engine components, and other critical parts that experience extreme temperatures during flight or re-entry. Compositions of geopolymer composites may be tailored to have a special thermal expansion coefficient, allowing them to be utilized for applications requiring dimensional stability at changing temperatures.

   Geopolymer concrete is used for the stabilization and immobilization of THM (toxic heavy metals) to reduce the impact on the environment when they are found in mining tailings management or acid mine drainage treatment. As a mineral that can resist acid attack and immobilize heavy metals, thus providing long-term storage or treatment of mining waste, it may also mitigate the environmental risks related to abandoned mines and active mining operations. Geopolymer-based materials form an impermeable barrier on mining sites, preventing the migration of leached contaminated groundwater.

2. **3D printing**: The customizable rheological characteristics of geopolymer concrete allow for the production of complex structures, assisting in better geometries and optimized designs that are difficult to produce with conventional concrete. These properties give designers more control over the printing process through adjustable setting time and flow characteristics of the material, leading to innovative architectural designs and custom construction solutions. The use of geopolymer in 3D printing can greatly reduce material wastage and labor costs for construction while enabling optimized designs that minimize material consumption and maximize structural performance.

3. **Repair and rehabilitation of aging infrastructure**: Geopolymer concrete, with its superior bonding characteristics and compatibility with existing structures, can provide affordable solutions for extending the service life of bridges, buildings, and other critical assets. The high-strength and low shrinkage properties of the material enable it to be used effectively in patch repairs and structural strengthening applications, allowing for quick curing and minimizing downtime/disruption during rehabilitation works. Repair materials based on geopolymer can also be designed to match the color of existing concrete, allowing for aesthetic integration in repaired structures.

4. **Energy sector**: Geothermal wells and nuclear power plants require materials that can withstand very high temperatures, radiation, and other extreme conditions in geo-environmental applications. Geopolymer concrete has proven long-term stability under such aggressive operating conditions. Its high toughness ensures superior operational efficiency and lower maintenance in such harsh environments, thus increasing the safety and reliability of energy generation plants. The low thermal conductivity of geopolymer concrete is also advantageous for insulation applications in energy infrastructure, improving efficiency and avoiding heat loss.

5. **Agriculture**: The chemical resistance to organic acids provided by geopolymer concrete makes it suitable for use in animal waste management systems and storage structures, thus extending the service life of agricultural structures. It is extremely beneficial against exposure to corrosive elements that are commonplace in manure storage tanks, silage bunkers, and composting spots. This means that the utilization of geopolymer concrete in these applications can minimize maintenance expenditure and any environmental risks due to structural failure. Geopolymer-based permeable reactive barriers and other applications in treating agricultural runoff will also reduce nutrient loading to waterways.

6. **Pavements**: Geopolymer is beneficial in non-aqueous situations where maintenance is expensive or harmful (such as transportation infrastructure, including railway sleepers and airport runways) because to its improved freeze-thaw cycle and de-icing salt resistance. The material's improved flexural strength and fatigue resistance make it ideal for high-density areas that experience frequent stresses, such as bridge decks and highway pavements. These characteristics lead to increased durability and lower life cycle costs for transportation infrastructure. Geopolymer concrete also gains high early strength, ideal for reopening repaired roads and runways more quickly than ordinary concrete, so disruption of the transportation network is minimized.

This opens up applications in the restoration of historical monuments and archeological sites, where geopolymer-based materials are compatible with ancient building techniques and materials, thus prolonging cultural artifacts. Geopolymer properties can be tailored according to the original materials desirable in a restoration project, providing seamless integration and maintaining the historical authenticity and structural integrity of ancient structures. Additionally, by using geopolymer, conservation materials may improve resilience to pollution and acid rain, two natural and man-made environmental degradations, better preserving our cultural legacy for future generations.

7. **Disaster-resistant construction**: By providing enhanced ductility and impact resistance for buildings constructed in seismically active or weather-prone locations, geopolymer concrete improves community resilience. This property of the material can absorb large amounts of energy and provide structural integrity despite severe loading for buildings and infrastructure, which improves the safety of structures located in high-hazard areas and may ultimately save lives

or reduce economic losses from natural disasters. Engineering self-healing properties in geopolymer concrete can also enable repaired structures to maintain their integrity from minor damage produced by seismic events or other natural forces.

Geopolymer concrete showing superior performance in submerged conditions makes it applicable as a material for underwater structures, including offshore wind turbine foundations, subsea pipes, and underwater tunnels. Because of its low permeability and resistance to chemical assault in the marine environment, it helps undersea constructions last longer and need less maintenance. The material's capacity to reach significant early strength underwater is a favorable attribute that makes it useful for building techniques like tremie concrete installation. Geopolymer concrete's resistance to biofouling may also lessen the requirement for protective coatings on coastal structures, which would have a smaller environmental effect and need less maintenance.

8. **Storage of thermal energy**: Concentrated solar power (CSP) plants and other thermal energy storage applications may benefit from the usage of geopolymer concrete because of its high thermal mass and resilience at high temperatures. By increasing efficiency and dependability, this material may be used as the foundation for centralized heat storage units that can effectively collect and release thermal energy, therefore participating in the renewable energy market. By permitting a steady energy supply from renewable sources, the integration of geopolymer with thermal storage strengthens sustainability by addressing the intermittent nature of solar and wind energy.

By adding conductive elements, such as metals, to geopolymer concrete, buildings with electromagnetic shielding may be created. These qualities are helpful in sensitive settings where protection from electromagnetic interference is sought, such as data centers, military installations, and medical institutions. They may be designed to reduce certain frequency bands, enabling shielding solutions that can be adjusted to meet individual demands for electromagnetic protection.

9. **Applications that reduce noise**: Geopolymer concrete is a potential material for noise barriers because of its density and the capacity to increase porosity by adding pore-forming raw components during manufacture. Additionally, the material may be utilized to construct sound barriers along railway tracks, roads, and industrial districts. In these places, lowering acoustic pollution is crucial to lowering environmental pollution levels in interface zones between cities and suburbs. The sustainability features of this alternative are further improved by geopolymer acoustic barriers, which enable the use of recyclable materials and support green infrastructure projects.

10. **Universe exploration**: Geopolymers, when prepared from local resources on different planets or moons, may be utilized as high-performing building materials for future extraterrestrial habitats. Geopolymer concrete's resistance to extreme environmental conditions, such as high temperatures, radiation, and vacuum, along with the potential use of local regolith as its raw material, makes

geopolymer concrete a promising technology for space colonization. It would be useful as it can be used to create structures that could shield inhabitants from radiation or use lunar materials for pressurized habitats or infrastructure components needed to sustain human life on other celestial bodies.

11. **Water treatment infrastructure**: Owing to its resistance to chemical attack and low permeability, geopolymer concrete can be used to construct water treatment infrastructure facilities, such as desalination plants and sewage systems. The ability of geopolymer concrete to withstand harsh environments can dramatically increase the lifespan of essential components in water infrastructure, decreasing maintenance costs and enhancing the reliability of water supply and sanitation systems. This property can also be used in designing innovative filtration systems or reactive barriers for groundwater remediation.

12. **Smart infrastructure**: Geopolymer concrete can embed sensors and conductive materials to create smart infrastructure capable of self-monitoring, delivering real-time data. By deploying this kind of technology, critical infrastructure assets like bridges, buildings, and roadways can be monitored and alerted to structural problems, resulting in preventive maintenance planning as well as better management of city assets overall. The smart geopolymer concrete structures can also play a vital role in making cities more resilient and efficient in line with the globally emerging contours of smart city initiatives.

13. **Geopolymer foams and insulating materials**: Besides their polymeric nature, lightweight geopolymer (or inorganic-organic hybrid) foams also possess superb thermal and acoustic insulation properties, similar to conventional petroleum-based insulating materials. They can be incorporated into building envelopes, industrial equipment, and transportation vehicles to improve energy efficiency and reduce noise pollution. In addition to excellent safety performance as insulation materials, geopolymer foams are also fire-resistant.

14. **Catalytic surfaces**: Geopolymer materials can be designed or modified with catalytic elements to produce surfaces that decompose air pollutants and organic contaminants. They may be used to urban water treatment systems, air-purifying pavements, and self-cleaning building facades. As land quality and attributes improved, sustainable development gained momentum.

15. **Constructing under harsh conditions**: Geopolymer concrete is perfect for special construction projects that take place in hostile environments, like deep-sea installations, arctic research stations, and containers for industries that deal with corrosive products, because of its resistance to high temperatures and harsh chemical conditions. This property can enhance safety and minimize maintenance in extreme environments because it will preserve its integrity even under severe conditions. Geopolymer concrete, owing to its multidimensional nature, finds applications in emerging sectors of engineering that can pave the road toward sustainable development goals.

3.6 Summary

Geopolymer concrete (GPC) possesses several benefits compared to conventional concrete in terms of acid attack resistance, early strength gain, low creep, and moderate drying shrinkage. GPC is mainly applicable for precast; so far, GPC has been implemented in railway sleepers, sewage mines, box culverts, bridge decks, and wall panels. On the other hand, GPC also has disadvantages, such as requiring toxic chemicals for activation and being sensitive to the reaction conditions during polymerization. GPC has been demonstrated in broad applications across a wide array of sectors, including civil engineering, construction, environmental remediation, aerospace, mining, 3D printing, and repair and rehabilitation sectors; energy sector; agriculture; transportation infrastructure; preservation of cultural heritage; disaster-resistant construction; underwater construction facilities; thermal storage (thermal batteries) systems; electromagnetic shielding; dimensional stability; thermal product systems; protective barrier applications; and acoustic barriers in buildings and structures; biomedical applications; biodegradable geopolymer; degradable composite; space concrete; foam; water treatment infrastructure; smart metering; smart ideas; and socio-geopolitical off-peak geopolymer development; foam orthosis panels; symbiotic ceramic carbon fiber geopolymers; geopolymer foams; polyester cement; sustainable pavements; geopolymer insulation; clear catalysis; surface activation; design in extreme environments; and bathtubs.

3.7 Highlights

- Use of geopolymer concrete provides excellent acid attack resistance, early strength, and low shrinkage.
- It is used in infrastructure, marine structures, high-rise buildings, and precast elements.
- Geopolymer concrete is a complementary sustainable building material with a lower carbon footprint, promoting waste utilization.
- The use of geopolymer concrete serves as an excellent way to immobilize hazardous waste and for soil remediation.
- Aerospace, mining, 3D printing, and repair and rehabilitation are application areas of geopolymer concrete.
- Geopolymer concrete has applications in the energy sector, agriculture, transportation, and preservation of cultural heritage.
- The properties of geopolymer concrete are suitable for use as disaster-resistant housing materials, underwater structures, and adaptability for space exploration."

References

P. Duxson, J.L. Provis, G.C. Lukey, J.S.J. van Deventer, The role of inorganic polymer technology in developing green concrete. Cem. Concr. Res. **37**, 1590–1597 (2007a)

P. Duxson, A. Fernández-Jiménez, J.L. Provis, G.C. Lukey, A. Palomo, J.S.J. Van Deventer, Geopolymer technology: the current state of the art. J. Mater. Sci. **42**(9), 2917–2933 (2007b)

S. Song, D. Sohn, H.M. Jennings, T.O. Mason, Hydration of alkali-activated ground granulated blast furnace slag. J. Mater. Sci. **35**(1), 249–257 (2000)

Chapter 4
Manufacturing and Curing Methods of Geopolymer Concrete

4.1 Methods of Production for Geopolymer Concrete

Methods of producing geopolymer concrete are varied and differ from one another in their special features and applications. These approaches were developed to improve the production process, mechanical properties of materials, and construction needs. Here is an in-depth analysis of the types of production and curing methods are:

1. Alkali activation: The process begins by mixing aluminosilicate materials (such as fly ash or metakaolin) with alkaline activators (typically sodium hydroxide and sodium silicate).
2. Preparation: Raw materials are ground and sieved to achieve the desired particle size distribution for optimal reactivity.
3. Mixing: Aggregates, cementetious material, and alkaline solution are combined in a mixer to form a homogeneous mixture.
4. Curing: The fresh geopolymer concrete is subjected to heat curing (usually between 40-80° C) for several hours to accelerate the geopolymerization process.
5. Ambient curing: Some geopolymer mixes can be cured at room temperature, though this may result in slower strength development.
6. Steam curing: An alternative method involving exposure to steam at elevated temperatures to enhance strength and durability.
7. Microwave curing: A rapid curing technique using microwave energy to accelerate the geopolymerization reaction.
8. Two-stage mixing: A method where dry ingredients are pre-mixed before adding the alkaline solution to improve workability and strength.

These methods can be adapted based on the specific raw materials, desired properties, and application requirements of the geopolymer concrete.

K. K. Poloju and K. Srinivasu, *Geopolymer Concrete*,
SpringerBriefs in Applied Sciences and Technology,
https://doi.org/10.1007/978-981-96-2479-9_4

4.2 Materials Used for Geopolymer Concrete

The production of geopolymer concrete consists of, among others, these raw materials (natural modifiers) that contribute to the characteristics and performance properties of the final product quality.

This includes four constituent categories:

1. Alkaline activators
2. Aluminosilicate source materials
3. Aggregates
4. Other materials.

Each of these categories has a variety of elements that contribute to the characteristics of geopolymer concrete. Davidovits is the first company to use alkaline solutions as activators to turn industrial waste byproducts into binding material. A lot of historical background on the characteristics of the GPC is then given by Wang et al. (1995). Fly Ash-Based Geopolymer Concrete: Results indicative of excellent strength. The use of a higher alkaline solution concentration (Puertas et al. 2000) and high temperatures provides strength values close to 50 MPa. Cement is replaced with source materials; in this case, the two types of source materials used are fly ash and GGBS. The basis for polymerization is the alkaline solution.

According to Puertas et al. (2000), fly ash-based geopolymers have certain disadvantages despite their great performance, including quick setting, poor processing qualities, demanding curing needs, and poor outdoor performance. For fly ash-based geopolymers to attain early stability, high temperatures (60–90 °C) are also required. The samples may need to be tested in a heat curing lab or pen. But in real-world situations, we are unable to create such circumstances for ambitious endeavors. To eliminate these deficiencies, trials were made using alkali-activated slag. Only two variables are necessary to measure mix proportions: the concentration of the alkaline solution and the AL/FA ratio.

Introduction: With high-performance concrete, a new trend of study has focused on fly ash and GGBS. However, there is a lack of appropriate combinations of designs and recommendations for GPC with GGBS, as studies have shown that fly ash-based geopolymer concrete can replace high-performance conventional concrete (Manjunatha et al. 2014). As a result, the current work takes into consideration parameters such as binder type, binder content, alkaline/binder ratio, efficient curing, and proper conditions.

At the top level, pure GGBS, which is a waste disposal product of the steel industry used to replace cement, can make a significant impact on cost/concrete reduction and open new horizons for ecological concrete. GGBS has improved both fresh and hardened concrete properties, such as reduction in workability during heat of hydration, high long-term strength, improvement in corrosion resistance, reduction in porosity, and permeability, etc. Moreover, the introduction of fibers in the concrete can help to produce enhanced mechanical and durability properties.

4.3 Geopolymer Concrete Components

The physical qualities of raw materials needed to make geopolymer concrete are analyzed using Indian standards. Geopolymer concrete is a form of alkali-activated concrete because it includes aluminosilicate binders that need an alkaline solution to activate. When combined with this solution, it forms a binding paste that cure and hardens in 30–60 min, typically at room temperature. The mix typically consists of aggregates, fly ash, GGBS, and an alkaline solution. The geopolymer concrete's binding ingredients are fly ash and GGBS, and an alkaline solution is employed to allow these two elements to react and build up the polymerization process. Geopolymer concrete is typically fly ash-based, however GGBS may be used in varying quantities depending on the application.

1. **Binding material**: Aluminosilicate Source Materials are essential components of geopolymer concrete.

Fly ash: Fly ash is a byproduct of coal combustion and hence readily accessible. Because of its high silica and alumina content, especially the low calcium level of Class F fly ash, it has excellent control over early strength growth and setting time, making it a preferred variety among others.

Ground granulated blast furnace slag (GGBS): Ground Granulated Blast Furnace Slag is predominantly a cementitious material used in concrete that is produced as a byproduct of the steelmaking process in blast furnaces.

According to Wang et al. (1995), a ton of steel equals 0.13–0.2 tons of slag; by 2020, about 310 million to 380 million tons of steel slag and 180 million to 270 million tons of iron slag are predicted to be produced worldwide. Ground Granulated Blast Furnace Slag (GGBS) is primarily made up of magnesium oxide, calcium oxide, aluminum oxide, and silicon dioxide, all of which can be found in the cement. GGBS is used in concrete because of its cementitious properties. Molten iron slag is quenched in steam or water in blast furnaces to produce a granular glassy solid, which is then dried to produce a fine powder.

Alkaline activator solution: Alkali-activated material (AAM) is created by mixing aluminosilicates, such as powdered and granulated blast furnace slag and fly ash, with an alkaline substance in a low-temperature range, resulting in the formation of a binding paste. In geopolymer concrete, the solution is applied to powdered aluminosilicate constituents, causing a chemical process known as geopolymerization.

Concrete is composed of a glue that binds its biggest components, coarse and fine particles. This material improves the mechanical qualities of concrete because it has a high degree of volume stability and stability under various forms of erosion. In concrete applications, they are further classified as coarse and fine aggregates. Coarse aggregates are rough and big in texture, measuring more than 4.75 mm, and are usually made up of fragmented stone particles, also known as gravel. In contrast, fine aggregates are made up of tiny particles such as broken stones and sand that may pass through a 4.75 mm filter. It is a main element of concrete, with an aggregate percentage of 60–80%. The coarse aggregates operate as an inert filler in the concrete mix, while the fine aggregates aid to fill any gaps between the coarse aggregate particles.

Alkaline activators have a key role in both initiating and prolonging the geopolymerization process. One of the most well-known activators is sodium hydroxide (NaOH), which is a strong base and is often utilized. It combines aluminosilicate resources and aids in the formation of the geopolymer network. Sodium hydroxide is avoided for minimal efflorescence. Potassium hydroxide (KOH) is used instead of sodium hydroxide when it is desired to reduce the possibility of surface appearance issues. Sodium silicate (Na_2SiO_3), often known as water glass, is widely used with sodium hydroxide. This is because it adds more silica to the geopolymer network, which may improve durability and strength. Potassium silicate (K_2SiO_3) functions similarly to sodium silicate, except potassium is used instead of sodium, which may increase performance in specific mix designs. Various combinations of these activators are often employed to optimize the geopolymerization process and obtain certain product qualities.

Aggregates: The function of aggregates in geopolymer concrete determines volume stability, strength, and other mechanical properties.

Coarse aggregates, such as crushed stone or gravel, comprise the majority of concrete and give significant compressive strength. Coarse aggregates, on the other hand, have an important role in defining the workability of new concrete and the binding strength between aggregate and geopolymer paste due to its size, shape, and texture.

Fine aggregates, often sand or crushed rock fines, are used to fill the spaces between coarse aggregates, increasing density and decreasing porosity in concrete. Gradation affects the packing and workability of fine aggregates. Recycled concrete aggregates are made from broken concrete collected at demolition sites and may serve as an ecologically friendly replacement for natural materials. The use of alternative aggregates may reduce the environmental imprint associated with concrete manufacturing, playing an important role in promoting a circular economy in building. Lightweight aggregates such as expanded clay, shale, or slate may be utilized to create concrete with the appropriate lightweight and thermal insulation qualities. They may also be used in high-rise buildings or other applications where weight reduction is critical.

Superplasticizer: A naphthalene formaldehyde sulfonate superplasticizer is used to increase workability. IS 9103 describes a superplasticizer (SP) as a high-range water reducer intended to improve concrete workability. High-performance sulfated naphthalene formaldehyde is employed. 0.5% superplasticizer by weight of binding substance. Other materials include different additions and components that may alter the qualities of geopolymer concrete. Water is the most abundant component when employed to provide adequate workability and the required geopolymerization process.

The water-to-binder ratio is modified to strike an appropriate balance between workability, strength, and durability requirements. Superplasticizers, also known as high-range water reducers, are a kind of chemical additive that improves and maintains the workability of concrete mixes without the need for extra water. It also permits the creation of low-water, high-strength geopolymer concrete.

Advanced geopolymer concrete technique incorporates nanomaterials such as nanosilica and carbon nanotubes. Some of these compounds may also increase concrete quality by increasing their strength, longevity, or even making it self-sense. Calcium-based additives have been employed in certain circumstances to modify the setting times and early strength development of final mixes, particularly those based on low-calcium precursors like Class F fly ash.

Specific combinations and amounts of these components are determined based on the desired properties of the finished geopolymer concrete product. The mix design is influenced by factors such as strength requirements, environmental exposure circumstances, and special application demands. A geopolymer concrete for a maritime project is often developed for durability and chloride intrusion resistance, while a precast mix may concentrate on strength increase and shrinkage.

Geopolymerization is a process that begins with the dissolving of aluminosilicate materials in an alkaline solution and ends with the development of a three-dimensional aluminosilicate network. It is influenced by curing temperature, curing duration, and the chemical makeup of the precursor components. In comparison to ordinary Portland cement, the resulting geopolymer binder exhibits improved acid resistance, fire resistance, and alkali-silica reactivity.

To summarize, the formation of geopolymer concrete is a complex interaction of its many components, which results in geopolymers and their performance. This customization has the potential to provide long-lasting, high-performance concrete solutions for a broad range of building and infrastructure applications. With continued research in this field, it is envisaged that new materials and processes will be developed to expand the uses of geopolymer concrete.

4.4 Preparation of Geopolymer Paste

In a pan mixer, the dry fly ash and GGBS are blended in various proportions. For three minutes, combine the blend with the alkaline solution to bring it into a consistent state. Different percentages of source material and sodium hydroxide concentration are contained in the various geopolymer pastes with variable quantities of alkaline activators prepared.

4.5   Curing Methods

Ambient curing: Here, raw materials are mixed and treated at room temperature. This, therefore, allows for a wide range of construction scenarios—ideal for precast and on-site applications. Now, the method has a lower rate of strength gain compared to others, since heat curing methods provide quicker and earlier strength gain; thus, this method can be beneficial in situations where a stepwise increase in strength is desired. Ambient curing has a significant advantage by generating less energy and carbon footprint; hence, it can be considered an eco-friendly process. The extended time for curing means that project timelines may need to be adjusted, and planning needs to be done accordingly. Moreover, ambient curing requires less discipline in construction scheduling because it doesn't involve specialized equipment or controlled environments. This technique is extremely useful in remote areas where access to sophisticated curing facilities may be challenging.

Additionally, certain geopolymer formulations have shown reduced shrinkage and improved durability, maybe as a result of slower reaction rates that permit deeper geopolymerization processes.

Heat curing: In this technique, the geopolymer mixture is heated to a high temperature of around 60–90 °C for a number of hours. Thus, this method strengthens the concrete faster and improves its mechanical properties. Heat curing is generally used in precast plants where controlled conditions can be easily provided. Because of the rapid curing process, we get faster curing cycles and the ability to use concrete elements on-site. Still, heat curing consumes more energy than ambient curing and may lead to higher production costs and environmental impacts. Heat curing employs higher temperatures, but that speed comes at the cost of rapid moisture loss, which, if prolonged, can cause micro-cracking and reduced durability in the long run. In order to reduce such problems, it is necessary to control the temperature ramp rates and humidity at which curing occurs. For example, geopolymer concrete mixtures that have high activation energy in order to initiate the geopolymerization reaction (due, for example, to the type of fly ash or metakaolin used) could particularly benefit from heat curing.

Steam curing: Like heat curing, steam curing uses higher temperatures that induce curing in a shorter time. However, steam is used, creating a humid environment during the curing period. This approach helps increase the development of strength and will also reduce shrinkage, increasing both durability and dimensional stability of the concrete. For high-volume precast production with a consistent climate across major components, steam curing works very well. By producing saturated moisture vapor, it prevents moisture loss and initiates geopolymerization processes. Steam curing provides many benefits over dry heat curing, including improving the pore structure, lowering the chance of surface drying and cracking, and distributing the warmth more evenly throughout the concrete mass. Steam also helps transport any alkali activator that will assist in geopolymerization throughout the matrix, which can enhance the process. Although steam curing (also known as wet curing) is used, it requires special equipment and strict control of the temperature and pressure parameters for optimal results. Also, when calculating the ecological footprint of this method, one should bear in mind that steam generation requires energy.

Oven curing: This novel technique uses microwave radiation to speed up the geopolymerization process. The curing process takes a significant amount of time with conventional methods, but microwave curing has quick strength development potential (microwave curing is three times faster). Another innovative technique that can lower manufacturing costs and environmental effects is heat curing, which uses less energy than other conventional heat curing techniques. Only small-scale manufacturing microwave equipment can use the microwave curing process, and it is still difficult to guarantee consistent heating of massive concrete components. Microwaves generate heat through a different mechanism than conventional heating, essentially by directly transferring energy to the material, allowing for more effective curing processes than conventional heating methods. It can yield positive microstructure evolution and mechanical performance. Moreover, microwave curing provides an opportunity to selectively heat phase(s) in the geopolymer mixture, permitting better-controlled reaction kinetics. Despite its potential advantages, large-scale microwave curing faces issues such as the need for special equipment that can treat large volumes of fresh concrete while providing uniform energy distribution and safety concerns regarding high-power microwave systems.

Two-part mixing: This involves separately mixing the activator solution and dry components before combining them to create the final geopolymer concrete mix. This means that the kinetics of the reaction should be well controlled and tailored with workability in mind to accommodate various applications. From precast to on-site, two-part mixing also provides flexibility for a range of construction applications. However, it requires proper handling of alkaline activators that can be hazardous and will necessitate skilled training for workers. There are many advantages to this two-part mixing approach regarding the flexibility of mix design and quality control. Separating the activator solution until just before application gives the individual components a long shelf life and reduces the risk of premature reaction.

This, in turn, allows fine-tuning of the activator concentration and composition for every reaction based on the nature of the source material or the desired properties of its geopolymerization product. Nevertheless, liquid alkaline activators are not suitable for on-site applications due to issues with storage, transportation, and handling. It is important to ensure that there are necessary safety protocols and gear to protect workers from potential chemical exposure. Furthermore, the two-stage mixing process might require more specialized gear and a greater degree of technical understanding than standard concrete mixing.

One-part mixing (just-add-water): The process itself is simplified by pre-mixing all dry components, including activators. It is, however, mixed on-site with water, exactly as is done when using conventional concrete. One-part mixing has several benefits, especially easy handling and lower safety issues involved with liquid alkaline activators. This process is particularly useful for on-site usage, where storage or handling of multiple components may be difficult. Still, one-part geopolymer blends come with a relatively limited stability period as their dry constituents might be reactive and thus might require special storage conditions and careful handling skills. Many practical concerns about the use of this kind of geopolymer concrete in conventional construction techniques are addressed by this one-part mixing procedure. By providing a product that can be mixed and laid using tools and techniques that are familiar to contractors and laborers used to conventional concrete, it lowers the entrance barrier. What this approach also does is mitigate the logistical challenges that come with transporting various components around construction sites and storing them. Nonetheless, there are major technical issues that must be addressed to develop effective one-part geopolymer mixtures. The dry activators must be appropriately selected, and proportions should be determined so that they dissolve properly and readily react with each other when water is added. The blend needs to allow for proper working time but also to ensure that the required strength development takes place. Much work is currently being done to stabilize one-part geopolymer systems for longer-term shelf life and performance and to enhance the number of source materials that could be effectively employed in this manner.

Hybrid alkaline-cement modes: These include combinations of geopolymer binders and a small content of ordinary Portland cement (OPC). OPC also contributes to early strength gain and workability; thus, it overcomes some disadvantages of pure geopolymer systems. Hybrid systems strike a balance between the environmental advantages of geopolymers and the known characteristics of customary cement-based concrete. This method offers a more sustainable alternative without sacrificing handling qualities, making the switch from conventional concrete to geopolymer concrete in construction seamless. When utilized in hybrid systems, this enables the resilience associated with OPC, including improved setting properties, reduced vulnerability to curing conditions, and increased compatibility with conventional concrete admixtures. The calcium obtained from the OPC will help create more binding phases that have strong durability and mechanical qualities. To achieve the necessary performance qualities without negating any environmental advantages, a precise balance between the quantity of geopolymer and OPC must be struck. The intricate relationships between geopolymer and cement hydration products, as well as certain combination mix designs to optimize appropriate performance for various scenarios, are currently being studied.

Depending on the application, resources available, environmental factors, and project requirements, each of these manufacturing processes has advantages and disadvantages of its own. With many ongoing studies of geopolymer concrete, it is expected that the production methods will continue to evolve and improve, as well as research on its uses. The development of geopolymer concrete production methods is still ongoing and driven by several key factors, including:

Sustainability: With the construction segment under scrutiny to limit its ecological footprint, there has been an increasing effort to improve geopolymer technology for high-energy efficiencies, low carbon emissions, and efficient reuse of wastes.

Improved properties: Using various mixing techniques, curing methods, and activators, ongoing research at labs and factories attempts to enhance the mechanical characteristics, durability, and long-term effectiveness of geopolymer concrete.

Scalability: In order to produce geopolymer concrete at industrial scales, production techniques must have the same qualities and capabilities as regular (Portland) cement concrete, producing material of equal or higher quality.

Standardization: Established production methods and quality control for geopolymer concrete are in the advanced stage, which will help increase the use of this technology in a wide array of the construction industry with consistent performance.

Cost-effectiveness: Continued research seeks to improve geopolymer concrete production methods to decrease raw material, processing, and curing costs, improving the economics of geopolymer concrete relative to traditional concrete.

Versatility: Producing methods will be developed to allow a wider range of source materials and activator types, increasing its potential applications and ability to fit local resources.

Ease of use: New production methods such as one-part mixing systems and better handling properties can help make geopolymer concrete more accessible to conventional concrete practitioners.

Future advancements in geopolymer concrete production are driven by these factors, and it can be believed that new methods, along with hybrid methods, will come into existence with many options available for drivers of manufacturing and construction. This continuous change in production methods is one of the key factors for a more rapid implementation and adoption of geopolymer concrete as an environmentally friendly substitute for cement-based construction materials.

4.6 Tests on Geopolymer Paste

Normal consistency: As represented in Fig. 4.1, the consistency of fly ash and GGBS paste is measured by Vicat's technique. Instead of regular water, an alkaline solution is used. Results of the standard consistency test are the same as those of cement regular consistency testing. This is because fly ash and GGBS form a common geopolymer paste within an alkaline solution. The consistency of geopolymer paste of normal was measured as the proportion of alkaline activator penetrating Vicat's plunger to a depth of 33–35 mm from the top face of Vicat's mold. As per IS 4031-1988 (1988), the normal consistency has been allowed to penetrate in Vicat's with a plunger of 10 mm diameter to a depth of 5–7 mm from the bottom of Vicat's mold. To determine this, trial pastes with different percentages of alkaline solution and GGBS are prepared and tested until the amount of alkaline solution required to give normal consistency is found.

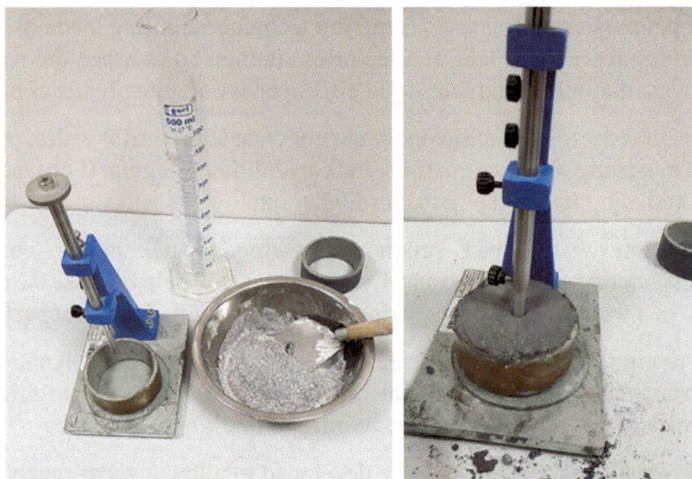

Fig. 4.1 Preparation of geopolymer paste to determine normal consistency

Setting time: The alkaline activator solution is manually mixed with the fly ash in a bowl to ensure reactivity. The fly ash and GGBS based geopolymer paste are made right before being mixed with the fly ash. Instead of cement and water, fly ash and GGBS with an alkaline solution are employed in the cement paste requirements. Vicat's apparatus is also used to calculate the ultimate setting time for a combination of fly ash and GGBS paste. An alkaline activator is required 0.85 times more in a standard geopolymer paste. The alkaline solution is added to the fly ash and GGBS during the final setting time and a 1 mm needle leaves an imprint on the paste in the mold but does not appear with a 5 mm attachment. The amount of AAS utilized impacts how long the geopolymer paste takes to cure (Alkaline Activated Solution). According to IS 4031-1988 (1988), the setup time is assessed using the Vicat device. Scanning Electron Microscopy (SEM) is a kind of electron microscopy. The microstructure of fly ash and GGBS may be studied using scanning electron microscopy (SEM). The fly ash particles are spherical and mainly composed of silica and alumina.

Test configuration and procedure

The Vicat's apparatus is used to determine cement's average or standard consistency, according to IS: 4031 (Part IV) (1988). A similar technique tests geopolymer material; an alkaline solution generates standard consistency geopolymer paste. The normal consistency of the geopolymer paste is defined as the amount of alkaline activator that allows the plunger of the Vicat apparatus to penetrate to a depth of 33–35 mm from the top of Vicat's mold. Vicat's apparatus, 400 g of fly ash and GGBS based binder combinations, and 0.85 times of alkaline activator are used to

Fig. 4.2 Preparation of geopolymer mortar to determine compressive strength

compute the geopolymer paste's final setting time (0.85 P). Three 100 mm diameter geopolymer mortar cubes from each set are cast and squeezed for 28 days to determine compressive strength.

4.7   Tests on Geopolymer Mortar

The Compressive Strength of Geopolymer Mortar

The compressive strength ratios of the raw materials Fly ash + GGBS, and alkali binder is 1:1. In this experiment, the mixture of both is kept constant. The specimens cast are shown in Fig. 4.2.

4.8   Workability Test on Geopolymer Concrete

Workability refers to how easily newly mixed concrete may be combined, transported, placed and compacted to avoid homogeneity loss or segregation. Because it directly affects the caliber and strength of the concrete material, it is a crucial attribute to be checked. It may also be described as the amount of internal labor necessary to compress concrete completely. As a result, it may be claimed that compaction and the workability attribute are connected. Each grade of concrete is given following the needs of structural design and construction. Thus, the desired workability of each stage varies. The strength needed for more extensive projects necessitates a higher grade of concrete, yet a high strength requires less workability. For instance, a high degree of workability is recommended for a substantially reinforced element or a small, inaccessible piece of the structure. When the alkaline solution is used in place of water in geopolymer concrete, workability is categorized based on the quantity of water in the concrete. Unworkable concrete is rough and has less water in it. Lower workability is achieved by having an alkaline solution-binder ratio of less than 0.4 (Fig. 4.3).

By contrast, medium workable concrete is defined as being easily mixed and compacted without experiencing significant bleeding or homogeneity loss and having a water-cement ratio between 0.4 and 0.5. The third type, high workability, designates plain concrete to mix, move, and compress. Without any effort, the material flows and settles down smoothly. The alkaline-binder solution would be more than 0.5 (Das et al. 2021). Workability may be determined using various techniques, including the Kelly ball test, Slump cone test, flow test, compaction factor test, and Vee-Bee consistometer test (Hemali 2019). This research will evaluate the workability of geopolymer concrete using the slump cone test. The slump cone test is the most used since it is inexpensive and yields findings immediately. To do this, a cone-shaped mold filled with oil is used, and the concrete mixture is progressively placed in stages, tamped down between additions. The form of the mixture is then represented by raising the mold, and the height difference between the liquid and the mold indicates the consistency of the combination. This approach is preferable for highly workable concrete.

Fig. 4.3 Workability test using slump cone apparatus

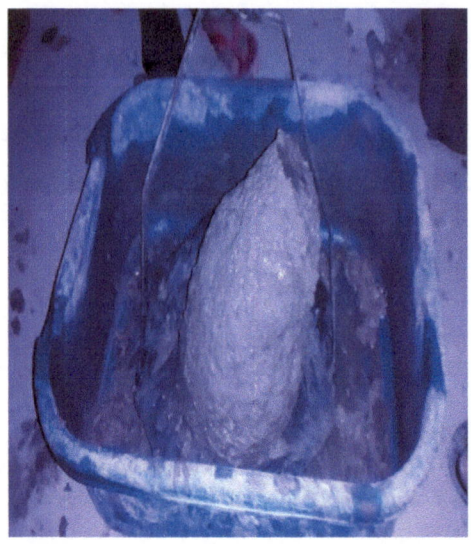

4.9   Mechanical Properties of Concrete

Concrete's compressive strength is crucial since it measures the material's capacity to withstand compressive loads. Finding the concrete's compressive strength is vital since it gives a general idea of whether the material is appropriate for a specific project and indicates its quality discussed by Phul (2019). Several elements, including aggregate quality, binding material quality, aggregate size, grading, and form, impact the compressive strength of the concrete. In OPC, the water-cement ratio may significantly affect the strength, and a greater strength can be reached by lowering this ratio. Still, in GPC, the alkaline solution-binder ratio, sodium hydroxide molarity, and sodium hydroxide-sodium silicate ratio can all impact the strength. Of course, the grade of concrete also matters since it directly affects compressive strength. The curing processes are also crucial; for instance, OPC is cured in water, whereas GPC is cured in heat or ambient temperature studied by Manickavasagam and Mohankumar (2017). These differences in curing processes and curing times can affect compressive strength in different ways; for example, 16% of the strength is attained in a day, whereas 99% is gained in 28 days. The grade of concrete, which is chosen based on the project's need for strength, is one of the crucial variables.

A grade of M50 should be used when aiming for a strength of 50 MPa owing to the mixed proportions that may provide this strength increment. GPC relies on aggregates, fly ash and any other binding substance as well as the alkaline solution and OPC depends on the quantity of aggregates, cement and water. The cubes utilized in this research will be tested in compliance with BS 1881-116, and the mechanical test will be conducted using a compressing testing machine in which a cube, cylinder, and prism concrete sample are inserted. Additionally, the cube is crushed as the device progressively increases stress from both sides until a failure happens due to compressive forces. By dividing the applied force at which the material failed by the cube's cross-sectional area, one may manually determine the compressive strength.

The Compressive Strength of Concrete

Cube specimens of size 150 mm × 150 mm × 150 mm are tested per IS 516-1956 (1956) on a compressive testing machine with a 2000 kN capacity, with the load applied on the cube specimen's casting face in the center and on the opposite sides after the required period of curing at ambient temperature in 7 and 28 respectively. The specimen is loaded at a constant rate of 140 kg/cm^2/min until it breaks down, at which point the highest load is regarded as the failure load. The strength of that combination is assumed to be represented by the average values of three specimens; individual variations should not exceed 15% of the average.

$$F_c = P/A$$

where

F_c is the concrete's compressive strength
P = Maximum load in N or kN at failure

A is the specimen's contact area in square millimeters (Fig. 4.4).

Split Tensile Strength of Geopolymer Concrete

Cylindrical specimens with dimensions of 100 mm in diameter and 300 mm in length are tested following the specifications in IS 516-1956 (1956) for split tensile strength. For the load to be applied along the 300 mm length, specimens are positioned horizontally in the machine between the two parallel steel strips, one at the top and one at the bottom. The formula given below is used to determine the specimen's split tensile strength after recording the highest load.

$$F_{\text{split}} = 2P/dl$$

F_{split} stands for concrete's split tensile strength.
$P =$ Maximum load at failure expressed in N or kN.
$d =$ The cylindrical specimen's diameter in millimeters.
l is the cylindrical specimen's length in millimeters (Fig. 4.5).

Flexural Strength of Geopolymer Concrete

The flexural strength of concrete acts a vital role in the design of structures. Flexural strength is expressed in the modulus of rupture, which is the maximum stress at the extreme fibers in bending. The size of the prism is 100 mm × 100 mm × 500 mm. The specimens are used to test the flexural strength of geopolymer concrete by bending them at two points in a 20 kN universal testing machine. The specimens are arranged in the machine so that the load will be applied along two lines 133 mm apart from the highest surface cast in the mold. The weight is gradually raised without applying any impact until the specimen fails. The test is done as per IS 516-1956 (1956). First,

Fig. 4.4 Compressive test of cube specimen

Fig. 4.5 Testing of a cylinder specimen for split tensile strength

the load is increased until the specimen fails, and the maximum load is noted as the failure load. The average of three samples is considered flexural strength.

For $a > 133$ mm

$$\text{Flexural strength } F_{\text{cr}} = Pl/(bd^2)$$

For 133 mm $> a > 110$ mm

$$\text{Flexural strength } F_{\text{cr}} = 3Pa/(bd^2)$$

where

P = maximum load applied to the specimen in N or kN.
l = Length of the span on which the specimen is supported in 'mm.'
b = Measured width of the specimen in mm.
a = Distance between a line of fracture and the nearer support measured on the centerline of the tension side of a specimen in mm.
d = Measured depth of the specimen in mm (Fig. 4.6).

Fig. 4.6 Testing of prism to determine the flexural strength

References

BIS (Bureau of Indian Standards), *IS 516-1956 (Reaffirmed 1999): Indian Standard Methods of Tests for Strength of Concrete* (Bureau of Indian Standard, New Delhi, India, 1956)

BIS: 4031-1988 (Part 4), *Method of Physical Test for Hydraulic Cement—Determination of Standard Consistency of Cement Paste* (Bureau of Indian Standards, New Delhi, 1988)

BIS: 4031-1988 (Part 5), *Method of Physical Test for Hydraulic Cement: Determination of Initial and Final Setting Time* (Bureau of Indian Standards, New Delhi, 1988)

S.K. Das, J. Mishra, S.M. Mustakim, An overview of current research trends in geopolymer concrete. Int. Res. J. Eng. Technol. **5**(11), 376 (2018). ISSN 2395-0056

P. Hemali, *Fly Ash for Concrete: Properties, Uses, Advantages & Disadvantages*. gharpedia.com (2019)

R. Manickavasagam, G. Mohankumar, Durability studies on the high calcium flyash based GPC. Int. J. Eng. Sci. Technol. **9**(4), 1–9 (2017)

G.S. Manjunatha, Radhakrishna, K. Venugopal, S.V. Maruthi, Strength characteristics of open air cured geopolymer concrete. Trans. Indian Ceram. Soc. **73**(2), 149–156 (2014)

A. Phul, M. Jaffar, S.N.R. Shah, A. Sandhu, GGBS and fly ash effects compressive strength by partial replacement of cement concrete. Civ. Eng. J. (2019)

F. Puertas, S. Martínez-Ramírez, S. Alonso, T. Vázquez, Alkali-activated fly ash/slag cement: strength behavior and hydration products. Cem. Concr. Res. **30**(10), 1625–1632 (2000)

S.D. Wang, X.C. Pu, K.L. Scrivener, P.L. Pratt, Alkali-activated slag cement and concrete: a review of properties and problems. Adv. Cem. Res. **7**(27), 93–102 (1995)

Chapter 5
Physical Characteristics of Geopolymer Concrete

5.1 General

Geopolymer concrete is an alternative to Portland cement concrete that is green and sustainable and has properties that are increasingly attracting the construction industry. This smart concrete has several distinct characteristics that set it apart from regular concrete:

1. **Sources**: Fly ash or ground granulated blast furnace slag (GGBS) is the primary component in geopolymer concrete, activated by alkaline solutions. Typically, these solutions consist of potassium hydroxide or sodium hydroxide mixed with potassium silicate or sodium silicate. This method offers a sustainable solution for industrial waste while also lessening its negative effects on the environment.
2. **Benefits to the environment**: The fact that geopolymer concrete has a much less carbon footprint than regular Portland cement concrete is one of its main advantages. Traditional cement production represents nearly 8% of global CO_2 emissions, with geopolymer concrete capable of reducing it by a further 80%, thus offering promising potential in the fight against climate change and as part of sustainable construction materials.
3. **Durability**: Geopolymer concrete has excellent durability properties, especially in extreme conditions. It shows excellent resistance against acid attacks, sulfate attacks, and fire. The improved resilience translates to longer service lives of infrastructure, lower maintenance costs, and better performance in adverse environments. This construction characteristic not only reduces the time frame of construction but also enables earlier stripping of formwork, which can prove to be a great asset in terms of both time and cost savings for construction projects. The final compressive strength of geopolymer concrete can be the same as that of normal concrete or even greater, depending on the mix and curing conditions.

K. K. Poloju and K. Srinivasu, *Geopolymer Concrete*,
SpringerBriefs in Applied Sciences and Technology,
https://doi.org/10.1007/978-981-96-2479-9_5

4. **Low shrinkage and creep**: Geopolymer concrete has lower drying shrinkage and creep than ordinary concrete. This property leads to less cracking and better dimensional stability in the long term, which is ideal for mass use when applied on a large scale as a bond between structures and infrastructure works.

5. **High-temperature stability**: At higher temperatures, geopolymer concrete has shown to be stronger than regular concrete. Because of this feature, it may be used in highly hot situations or in structures that need to be more fire-resistant.

6. **Corrosive and chemical resistant**: The material has outstanding resistance to a variety of corrosive conditions and chemicals. This quality makes geopolymer concrete a suitable material for industrial plants, waste processing facilities, and the grinding of other objects exposed to aggressive chemical agents.

7. **Workability**: The workability of geopolymer concrete can be like that of conventional concrete. It can be designed for different workability via mix design and proper use of admixtures; therefore, it finds application in diverse construction techniques.

8. **Curing**: Geopolymer concrete often needs to be heat cured for maximum strength (unlike traditional concrete). Often, this is done by holding the concrete at high temperatures (60–80 °C) for a certain amount of time. However, research is ongoing to develop ambient curing methods that would improve flexibility and ease of use regarding its application in different site conditions.

9. **Versatility**: It may be used in a number of projects, including 3D printing, cast-in-place construction, and precast parts. Its special qualities and flexibility to a wide range of building techniques serve as an excellent representation of its versatility.

10. **Cost-effectiveness**: Because geopolymer concrete uses specialized ingredients and techniques, it may initially cost more than regular concrete, but over time, it may prove to be more cost-effective. Over the course of a structure's service life, its enhanced resistance to damage, reduced maintenance requirements, and utilization of waste materials can provide financial benefits.

Although geopolymer concrete provides numerous benefits, there are drawbacks as well. The absence of standardized mix design processes and the scarcity of long-term performance data are some significant disadvantages. The need for heat curing in many applications can complicate on-site building logistics. These shortcomings should be addressed in future, and there is still room for development.

1. It is important for engineers, architects, and construction professionals planning to use geopolymer concrete as an alternative building material to have a better understanding of these properties. With the growing focus on sustainability and innovation, geopolymer concrete is one solution that offers environmental benefits without sacrificing performance and will continue to attract interest from the construction industry.

2. Geopolymer concrete is a recent construction material that has several advantages over Portland cement concrete. Due to its mechanical, durability, and environmental properties, it is a promising alternative for building materials. This new material has gained a lot of attention in the past couple of years for its capability

to overcome some deficiencies of traditional concrete and provide improved performance in several ways.

5.2 Mechanical Properties

The strong early compressive strength: Within the first 24 h of curing, geopolymer concrete often exhibits strong early compressive strength, surpassing that of traditional concrete. The chemical processes that occur during geopolymerization, which create a thick, robust matrix, are primarily responsible for the rapid strength development. This increase in early strength gain is especially useful for precast concrete applications and fast-tracked construction.

Tensile strength: Typically, like or less than that of ordinary concrete, subject to mix design and curing. Its tensile strength can also be improved with fibers or other reinforcing materials, making geopolymer concrete suitable for applications where tensile forces are a concern (such as pavement structures).

Flexural strength: When compared to regular Portland cement concrete, geopolymer concrete often exhibits higher flexural strength. For concrete products like beams and slabs that are exposed to bending moments, this extra bending strength is advantageous. In some situations, improved flexural performance may lead to a more effective design that uses less material.

Elastic modulus: On par with or less than that of conventional concrete, which is useful in situations where more flexibility and elasticity are required. By simply altering the mix design and/or curing process, geopolymer concrete may be made to have the necessary elastic modulus, making it flexible enough to accommodate different project requirements. When dealing with rotating elastic loads that need to be controlled or in seismically active areas, this physical property's elasticity may be required.

Shrinkage: Geopolymer concrete seldom shows higher drying shrinkage volumes than regular concrete, which reduces the likelihood of breaking. Enhanced shrinkage reduction is due to the distinct composition and reactions in geopolymerization. This property is even more helpful in large structures or when dimensional stability is more critical.

5.3 Durability Characteristics

Chemical resistance: This material is extremely resistant to acid attacks, sulfate attacks, and other aggressive chemical environments, thus can be used in harsh atmospheric conditions. Geopolymer concrete has less calcium hydroxide (the susceptible phase of conventional concrete), and it has a dense microstructure, which results in significantly higher chemical resistance. It is this property that renders geopolymer concrete perfect for industrial facilities, wastewater treatment plants, and other corrosion-prone environments.

Fire resistance: It has better fire resistance than conventional concrete and can remain structurally sound at elevated temperatures. Geopolymer concrete shows exceptional thermal stability and can resist up to 800 °C with minimal loss of strength. This property can be especially beneficial for high-rise buildings, tunnels, and other infrastructure where fire is a major safety concern.

Freeze–thaw resistance: Offers a usually high level of resistance to freeze–thaw cycles, however this might vary depending on the mix design. The kind and amount of activators, air-entraining agents, consumption, and matrix porosity all affect how resistant geopolymer concrete is to freeze–thaw. With the utilization of sound mix design and curing procedures, we can achieve a geopolymer concrete with appropriate freeze–thaw durability for use in cool climates.

Corrosion resistance: Improved resistance to corrosion of reinforcement due to lower permeability and dense microstructure. The semi-permeable characteristics of geopolymer concrete prevent the penetration of chlorides and other corrosion agents, safeguarding embedded steel reinforcement. This characteristic is especially useful for marine environments, bridge structures, and all other applications exposed to aggressive environments.

Abrasion resistance: It generally has greater abrasion resistance compared to normal concrete, which allows for high wear use. The enhanced abrasion resistance of geopolymer concrete can be attributed to its dense microstructure and the strong chemical bonds developed during geopolymerization. This property makes it ideal for pavements, industrial floors, and other heavy traffic or abrasive areas.

Lower carbon footprint: Carbon emissions from the geopolymer concrete manufacturing process are significantly lower than those involved in Portland cement concrete production because the calcination of limestone is not required during production. Around 8% of the CO_2 emitted worldwide is due to producing conventional Portland cement. Due to the low-crystalline nature of this reactive intermediate, it can be used for making geopolymer concrete where alternative binders and industrial by-products will reduce emissions during the construction phase by up to 80%, thereby enhancing sustainable building materials.

Use of by-products: Frequently uses fly ash, slag, or other industrial waste forms, which reduces landfill disposal and contributes to the concepts of a circular economy. Utilizing such secondary raw materials means that less waste goes to landfills and there is less need for new virgin raw materials. This is consistent with sustainable development and conservation of resources; therefore, geopolymer concrete is an eco-friendly material for construction projects. After 10 years, the term geopolymer was used by Joseph Davidovits to describe synthetic rock-like materials that are synthesized from inorganic aluminosilicate materials such as ashes or clay.

Reduces energy usage: Its production typically uses less energy than that of standard cement preparation, which lowers energy expenses. Higher temperature kilns used to create Portland cement are unable to compete with the lower temperature geopolymerization method. This lower energy requirement therefore results in a decrease in the amount of fuel used to manufacture geopolymer concrete, which further lowers greenhouse gas emissions.

Reduced water consumption: Some geopolymer concrete mixes use less water than regular concrete, protecting valuable groundwater supplies. The kind and quantity of alkaline activator in geopolymer concrete often determines how much water is needed. In situations where there is competition for limited water sources, conservation strategies are further supported by workable geopolymer concrete produced by mix design optimization that maintains lower water-to-binder ratios.

Recyclability: To further reduce its environmental effect, geopolymer concrete may be recycled and utilized as aggregates to create fresh concrete mixtures. Because geopolymer concrete is readily recyclable, it supports the circular economy and extends the life of construction components. This land has been utilized to limit the amount of garbage that is disposed of during building and demolition, in addition to minimizing the need for virgin aggregates.

Nevertheless, the effectiveness of geopolymer concrete depends on the mix design as well as the raw ingredients and environmental factors. It is essential to choose the appropriate kind of precursors, activators, and ratios to achieve the required qualities and functionality. Furthermore, in order to achieve the qualities of geopolymer concrete, curing conditions including temperature and humidity are essential.

For geopolymer concrete to be implemented more effectively worldwide, a few other issues must be resolved, including standardization, quality assurance, and stakeholder education. Furthermore, certain areas lack sufficient long-term performance data and design codes to quickly integrate it into other applications. To overcome these problems and raise knowledge of geopolymer concrete technology, however, research is still being conducted.

Geopolymer concrete may become more and more important as the building sector continues to prioritize sustainability and looks for ways to lessen its influence on climate change. As an alternative to traditional Portland cement concrete, this unique combination of mechanical, durability, and environmental qualities has a great deal of promise for usage in a range of applications.

Porosity, density, strength, and resilience to a range of environmental factors. An innovative and environmentally responsible alternative to traditional Portland cement concrete is geopolymer concrete. Strength, density, porosity, and resilience to weather conditions are among of its qualities that make it a promising material for a variety of building applications. The capacity of this cutting-edge material to lower the carbon footprint connected with building activities has garnered a lot of attention in recent years.

Strength:

In general, geopolymer concrete has a stronger compressive strength than traditional concrete, which might provide it an advantage over comparable location. Strength increases rapidly, with the first 24 h after curing accounting for a significant portion of the ultimate strength. Precast concrete and circumstances requiring early formwork removal benefit greatly from such quick strength growth. Strength may be impacted by the following factors:

1. Precursor types (such as fly ash and metakaolin)
2. The alkaline activator's composition
3. Conditions for curing (temperature and time)
4. The ratio of silicon to aluminum in the geopolymer matrix.

The source materials have a significant impact on the ultimate strength of geopolymer concrete. Fly ash, one of the most prevalent byproducts worldwide and a consequence of burning pulverized coal, is often used because of its advantageous chemical characteristics and potassium-rich availability. Metakaolin, a byproduct of calcined kaolin clay, is another often utilized source material that usually yields geopolymers with excellent strength.

The nature and amount of these source materials significantly affect the mechanical behavior of concrete.

The process of geopolymerization starts only when an alkaline activator, commonly a mixture of sodium hydroxide and sodium silicate, is mixed with the precursors. These constituents can also be proportioned and mixed in unique ways, adjusting their concentration along with the ratio to develop greater stiffness. The same idea applies—greater concentration equates to stronger concrete, but one must balance workability and set time.

The strength of geopolymer concrete is significantly affected by the temperature and duration of curing; elevated temperature curing is certainly beneficial for geopolymer concrete but does not hold true for conventional concrete, where ambient temperature is preferable. Curing for 24–48 h at 60–80 °C has shown the best strength. On the other hand, ambient-cured geopolymer concretes have also gained attention from researchers aiming to develop greater feasibility regarding field applications.

Another important factor affecting strength is the silicon-to-aluminum ratio in the geopolymer matrix. It defines the three-dimensional aluminosilicate network structure in geopolymers that is responsible for their strength. Depending on the feedstock and desired properties, optimal ratios range from 1.5 to 2.5.

Depending on mix design and curing conditions, geopolymer concrete can have compressive strengths between 40 and 100 MPa. Some unique formulations have achieved strengths of more than 100 MPa, which are on par with standard concretes with good performance. Because of its great strength, it may be used in a wide range of structural applications, including infrastructure and low- and high-rise structures.

Density: Geopolymer concrete typically has a density of between 2200 and 2400 kg/m^3, which is comparable to regular concrete. Nevertheless, the density may be modified by altering the mix design or using lightweight mixes along with air-entraining agents. Such flexibility enables the manufacturing of a wide spectrum of geopolymer concretes, ranging from lightweight applications with densities as low as 1400 kg/ m³ to high-density shielding applications with densities exceeding 2600 kg/m³.

Density-Related Factors:

1. Source material types and proportions
2. Aggregate selection
3. Geopolymer matrix porosity
4. Curing conditions.

The density of geopolymer paste that serves as the binder in the concrete is affected by both the source material type and their ratios. For instance, metakaolin geopolymers are generally less dense than fly ash geopolymers due to particle shape and reactivity effects.

The ultimate density of geopolymer concrete is influenced by the aggregate selection mix. River sand and crushed stone are examples of common aggregates that may be used to make normal weight geopolymer concrete. Expanded clay, perlite, or recycled glass may be added to applications to increase lightweight while maintaining enough strength and density.

Density also depends on the porosity of the geopolymer matrix (influenced by the mix design and curing conditions) and overall density. Of course, lower density also means a more porous matrix, which may compromise strength and durability.

The geopolymerization reactions associated with geopolymers are affected by casting and curing conditions, including temperature and humidity. This implies that the density of geopolymer concrete is influenced by the rate and extent of reaction. Adequate curing contributes to the formation of a denser, better interconnected geopolymer framework, enhancing the density and strength of concrete.

Porosity: The pore structure of geopolymer concrete is typically less porous than that observed in conventional concrete, giving it superior durability. The geopolymer concrete pore structure is illustrated as follows:

1. Smaller pore sizes
2. Less connected pore networks
3. Lower total porosity.

The decreased porosity is due to the formation of a dense geopolymer gel and the good space-filling ability of the reaction product. The uniqueness of the pore structure can significantly affect the performance and durability of the material.

Geopolymer concretes are characterized by smaller pore sizes than conventional concrete because the processes of geopolymerization yield a much more homogeneous and finer microstructure than what is achievable with hydration products in conventional concrete. These smaller pores render the intrusion of deleterious materials more difficult, thus improving the resistance of concrete against various deterioration mechanisms.

This disconnected pore network is one of the significant reasons for the better impermeability of geopolymer concrete. In contrast to the continuous pore structure of conventional concrete with interconnected capillary pores, geopolymer concrete has a more tortuous and discontinuous pore structure, resulting in lower permeability than traditional Portland cement-based concretes. By being well compacted, this trait decreases the infiltration and thus penetration of aggressive agents (like sulfates), optimizing the performance of these materials.

Geopolymer gel is known to fill available spaces efficiently, leading to lower total porosity in mixed volume. The reduction in porosity results in greater strength, less permeability, and more resistance against environmental conditions.

The approach helps reduce porosity, which can be further minimized with improved mix design and curing conditions. Thus, several aspects like the water-to-binder ratio, type and concentration of alkaline activators, curing temperature, and time affect the final pore structure of geopolymer concrete. More advanced techniques, such adding nanoparticles or using superplasticizers, might increase performance by refining the pore structure and lowering porosity.

Environmental stress tolerance: Geopolymer concrete has a longer lifespan than traditional concrete because it is more resilient to a variety of environmental variables. One of the main causes of the resurgence of interest in this material for harsh and infrastructural applications is its enhanced endurance. This indicates that it can tolerate a range of environmental conditions, such as:

Acid resistance: Because of its aluminosilicate network structure, which is more stable than Portland cement hydration products that are high in calcium, geopolymer concrete exhibits exceptional resistance to acid assaults. This resilience is particularly helpful in acidic conditions, such those found in chemical processing facilities, wastewater treatment facilities, and certain agricultural uses.

The fundamental chemistry of geopolymers is responsible for this increased acid resistance. In contrast to Portland cement concrete, which depends on calcium-based substances that are vulnerable to acid attack, geopolymers create a durable three-dimensional aluminosilicate network. Geopolymer concrete exhibits low degradation rates in acid mediums and retains its structural integrity longer than normal concrete under acidic conditions.

Resistance to sulfates: Geopolymer concrete is highly resistant to sulfate attack, which affects conventional concrete containing calcium hydroxide and calcium silicate hydrate. Geopolymer systems exhibit markedly improved resistance to sulfate attack, preventing expansion, cracking, and loss of strength in Portland cement concrete. This resistance is especially important in areas with above-average sulfate concentrations (e.g., soil rich in gypsum, sulfate-rich groundwater, marine environments).

Geopolymer concrete is an exceptional option for foundations, underground structures, and marine infrastructure, where conventional concrete may degrade due to higher resistance to sulfate attack.

Resistance to chloride penetration: The low porosity and dense microstructure of geopolymer concrete ensure high resistance to chloride ion penetration, thus minimizing the corrosion of reinforcement in marine environments. This is essential for the service life of reinforced concrete in seawater or de-icing salts.

Several factors contribute to geopolymer concrete's resistance to chloride penetration:

A more polished pore structure that is smaller and disconnected.

The fact is that calcium hydroxide reacts with chlorides, producing highly soluble calcium chloride.

There is a good chance that the chlorides will be chemically bound inside the geopolymer matrix.

In comparison to traditional concrete, this results in less chloride intrusion and may extend the service life of reinforced structures exposed to chlorides.

Resistance to fire: Geopolymer concrete is appropriate for fire-resistant applications because, in contrast to traditional concrete, it retains its structural qualities at high temperatures (up to 800 °C). The improved flame resistance is attributed to the intrinsic thermal stability of the geopolymer matrix and the absence of chemically bonded water, which spalls at high temperatures in conventional concrete.

The benefits of the better fire resistance of geopolymer concrete include:

1. Enhanced safety of structures (e.g., buildings and infrastructure)
2. Increased fire protection
3. This property is applicable to high-risk areas, such as tunnels and factories.

Studies have shown that even after prolonged exposure to temperatures that would degrade regular concrete, geopolymer concrete may maintain a significant portion of its original strength. This characteristic gives geopolymer concrete a major edge in situations where fire resistance is essential.

On the other hand, if properly air-entrained and designed for freeze–thaw cycles, geopolymer concrete can demonstrate decent resistance to cycling, which is not a characteristic common in all types of mixes or curing conditions. The factors affecting the freeze–thaw resistance of geopolymer concrete are similar to those affecting conventional concrete:

1. Macro-pore structure of the matrix
2. Degree of saturation
3. Presence of air-entraining agents.

While some previously mentioned geopolymer concrete has shown good performance in freeze–thaw resistance, other formulations may require individual mix design details to achieve the optimal characteristics needed for cold climates. Increased effort is directed toward optimizing geopolymer formulations to improve their freeze–thaw durability for applications in areas prone to severe winter conditions.

Carbonation resistance: Compared with ordinary concrete, geopolymer concrete exhibits better performance in carbonation resistance because there is no calcium hydroxide retained in the matrix, and it has a dense microstructure. One of the most significant concerns in conventional concrete, namely carbonation that can cause reinforcement corrosion, is not as critical for geopolymer systems.

Several factors contribute to the enhanced resistance of geopolymer concrete against carbonation:

1. The fact is that calcium hydroxide, a main compound in conventional concrete, undergoes carbonation.
2. This dense microstructure with much lower porosity inhibits the penetration of CO_2.
3. The potential for chemical binding of CO_2 in the geopolymeric matrix.
4. Such features may enhance carbonation resistance rates and therefore provide a longer service life for structures subjected to CO_2-rich conditions.
5. Resistance to alkali-silica reaction (ASR): The availability of alkalis for ASR is limited in geopolymer concrete, making it less susceptible to this type of distress compared with ordinary Portland cement (OPC) concrete.
6. Unlike ASR, which causes expansion and cracking in conventional concrete, the same behavior is mitigated in geopolymer systems due to the binder's chemistry.
7. The resistance of geopolymer concrete against ASR can be related to the following factors:
8. The amount of alkalis consumed in the geopolymerization process, leading to fewer free alkalis that can react with aggregates.
9. Possible chemical binding of alkalis in the geopolymer matrix.
10. Relative to conventional concrete, these could include the different pore solution chemistry.
11. This negative reaction against ASR also allows for the use of more versatile aggregates in geopolymer concrete, which may demonstrate aggressive behavior in conventional concrete systems.
12. Geopolymer concrete shows better resistance to environmental factors, making it ideal for use in harsh environments, including marine structures, industrial floors, and wastewater treatment plants. The capacity of geopolymer concrete to endure harsh conditions can lead to lower maintenance needs and longer service lives for structures, which may compensate for the higher initial costs associated with producing this material.

High strength can be obtained with geopolymer concrete, but it comes at the cost of density, which is a favorable factor for certain applications. Moreover, geopolymer concrete also possesses low porosity and resistance to acid, sulfate, and seawater. These features—along with its significantly lower environmental footprint than traditional concrete—make it an attractive material for sustainable building. With ongoing research and evolving practical applications, geopolymer concrete is poised to offer substantial improvements in durability for construction without sacrificing sustainability.

Geopolymer concrete technology continues to advance and is concentrating on the following topics:

1. Mix design assessment for each application and local materials
2. Enhanced ambient temperature curing methods for better practicality on site
3. Essential testing and standard procedure development

4. Assessment of the long-term behavior and stability in different environmental conditions
5. The inclusion of geopolymer concrete in existing construction practices and standards
6. With the advancement of these research fields, geopolymer concrete is likely to be an increasingly better choice for sustainable and durable construction applications everywhere.

5.4 Summary

Geopolymer concrete is a new-age environmentally friendly alternative to traditional Portland cement concrete. Made using industrial by-products such as fly ash or slag upon activation with alkaline solutions, it comes with eco-friendly advantages like reduced CO_2 emissions. Geopolymer concrete possesses rapid strength development, high durability, low shrinkage and creep, thermal stability, and chemical resistance properties, as well as versatility in application. However, it is challenged by limited long-term performance data and, in some instances, a requirement for heat curing. Geopolymer concrete has better properties than OPC concrete; it exhibits higher early-age compressive strength compared to conventional concrete, whereas the tensile and flexural strengths are comparable, but with a lower elastic modulus. It is also known for its excellent durability properties, including resistance to chemical attack, fire, freeze–thaw cycles, and reinforcement corrosion. Geopolymer concrete is more environmentally friendly, having a smaller carbon footprint, using industrial waste, requiring less energy and water to produce, and can be recycled. Geopolymer concrete strength, density, porosity, and durability against environmental conditions are dictated by source materials (fly ash/binder composition), the nature of the alkaline activator, mix design, and the type of curing it underwent, etc. It can attain high compressive strengths, a density similar to that of traditional concrete, and lower porosity. Resistance to acid attack, sulfate attack, chloride penetration, fire, carbonation, and alkali-silica reaction is high in attained concrete. Future research is aimed at providing more optimal mix designs, identifying the best methods for ambient curing, creating industry-wide testing standards, and evaluating long-term performance to facilitate the widespread usage of this viable green building material.

5.5 Highlights

1. Geopolymer concrete consists of industrial wastes that are activated by alkaline solutions.
2. It has a much lower carbon footprint than Portland cement concrete.
3. Geopolymer concrete is extremely durable and performs well under extreme conditions.

4. It shows fast strength gain due to less water coagulation and good early strength.
5. Drying shrinkage and creep in geopolymer concrete are lower than those in ordinary concrete.
6. It shows better performance at elevated temperatures compared to normal concrete. Geopolymer concrete can be used for various purposes.

Chapter 6
Illustrations of Geopolymer Paste, Geopolymer Mortar, Geopolymer Concrete, and Mix Design Approach

This chapter discusses the different design and specification considerations for geopolymer concrete, including mix design, strength requirements, and other relevant design factors of M20 and M50 grade of GPC with an example. This includes components, testing methods etc.

6.1 Components of Geopolymer Concrete

According to Indian Standards, the physical characteristics of the components used to create geopolymer concrete are evaluated. Since concrete contains aluminosilicate binders, which cannot be activated without an alkaline solution, geopolymer concrete is an alkali-activated kind of concrete.

Once combined with this solution, a binding paste will be generated, set, and hardened in a short time, typically at room temperature. Aggregates, fly ash, GGBS, and the alkaline solution often make up the mix. The binding elements in geopolymer concrete are fly ash and GGBS, and the alkaline solution is utilized to allow these elements to react and create the polymerization process.

Typically, geopolymer concrete is based on fly ash, but depending on the situation, it may also include up to varied amounts of GGBS. The calcined materials had specific gravities of 2.2 and 2.9, respectively. However, Table 6.1 displays the chemical composition. Through SEM test for the source materials, it is discovered that the GGBS is angular in form and contains spherical fly ash particles containing silica.

K. K. Poloju and K. Srinivasu, *Geopolymer Concrete*,
SpringerBriefs in Applied Sciences and Technology,
https://doi.org/10.1007/978-981-96-2479-9_6

Table 6.1 Chemical composition of source material

Chemical composition	Fly ash (%)	GGBS (%)
Lime	04.00	32.50
Silica	60.10	34.00
Alumina	26.50	20.00
Iron oxide	04.25	00.80
Magnesia	01.25	07.89
Sulfur trioxide	00.35	00.90
Soda/potassium	00.22	Nil

6.1.1 Fly Ash and Their Types

Fly ash is a type of hazardous waste that is made up of extremely small ash particles that are released after the combustion of fuel, primarily coal.

The global fly ash market is estimated to be worth USD 6.99 billion in 2021 and is anticipated to increase to USD 12.46 billion by 2029, growing at a CAGR of 7.50% from 2022 to 2029.

India and China will produce over 112 million tons and 100 million tons of fly ash annually by the year 2022, respectively, with the US and Germany coming in second and third.

Due to the region's expanding demand from the cement and construction industries, it is expected that the Asia–Pacific region would have the largest fly ash market.

To enhance the characteristics, structure, and performance of concrete, flyash is utilized as a binding agent. This substance is created at electric generating facilities drawn out of pulverized coal during combustion. Therefore, when the coal burns, the mineral impurities like shale, quartz and clay fuse in suspension and float out of the burning chamber with the exhaust gases. Upon rising from the burning room, they cool and harden into fly ash like particles. Afterward, the material is collected using an electrostatic precipitator, a filter bag or both. According to Basham et al. (2007), these are the primary two forms of fly ash that are often used in concrete and will also be evaluated in this research for geopolymer concrete, the types of fly ash are classed based on their chemical makeup as shown in Figs. 6.1 and 6.2.

6.1.2 Fly Ash, Class F

The carbon concentrations range from less than 5 to 10% and this class is often rich in silicate and low in calcium. Bituminous and anthracite coals are used to make the ashes used to create class F. It has a higher pozzolanic content than class F, with pozzolanic compounds such as iron oxide, alumina oxide and silica oxide in amounts as high as 70%. Some have just a trace amount of calcium oxide, which

Fig. 6.1 Fly ash colors based on the mineral composition (Hemali 2019)

Fig. 6.2 Fly ash classes C and F (Hemali 2019)

may range from 8 to 16% of the substance. Due to its high pozzolanic content, this kind is resistant to chemical and sulfate assault, decreases reinforcement corrosion and experiences a less alkali-silica reaction. However, it has a low cementing value, or no cementing property mentioned by Boral Resources (2020).

6.1.3 Fly Ash, Class C

In contrast to class F, this kind has a high calcium concentration and a low silicate content. Its carbon content is often less than 2%. Additionally, they come from lignite and sub-bituminous burning coals. This kind is self-cementing and pozzolanic, albeit to a lesser extent than class F fly ash. Due to the high calcium content, the pozzolanic compounds range from 50 to 70% of the total composition. Free lime, tricalcium aluminate, calcium alumina sulfate glass and quartz comprise most of its design.

Fig. 6.3 Ground granulated
blast furnace slag

The variety is rich in calcium and low in silicate. To reduce various types of attacks and enhance concrete longevity, higher replacement percentages for this type of fly ash are advised by Boral Resources (2020).

6.1.4   Ground Granulated Blast Furnace Slag (GGBS)

Ground Granulated Blast Furnace Slag is a cementitious material that is mostly utilized in concrete, this substance is a by-product of the steel manufacturing process in blast furnaces.

A ton of steel produces 0.13–0.2 tons of slag as discussed by Yu and Wang (2011). Between 310 and 380 million tons of steel slag and 180 million–270 million tons of iron slag are projected to be produced globally in 2020.

Due to its cementitious qualities, ground granulated blast furnace slag is utilized in concrete. In blast furnaces, the molten iron slag is quenched in steam or water forming a granular glassy substance. In a subsequent step, this material is dried and turned into a fine powder as shown in Fig. 6.3. Magnesium oxide, calcium oxide, aluminum oxide and silicon dioxide are the critical ingredients of GGBS. Additionally, since these minerals are most often found in cementitious materials, they are the typical minerals that make up the cementing characteristic. Furthermore, these minerals provide GGBS with more strength since they include calcium silicate hydrates, a substance that improves strength CSMA (2021).

6.1.5   Alkaline Activator Solution

An alkaline solution is created when an alkali, such as sodium hydroxide, dissolves in water. This solution induces a chemical process called "Geopolymerisation" when applied to powdered aluminosilicate components in geopolymer concrete. When aluminosilicates like powdered, granulated blast furnace slag and fly ash are combined with an alkaline solution at room temperature, a cementitious binding paste is created. Sodium hydroxide, magnesium hydroxide, calcium hydroxide or potassium hydroxide may be used to develop alkaline solutions. However, either potassium

Fig. 6.4 NaOH pellets

Fig. 6.5 Na$_2$SiO$_3$ solution

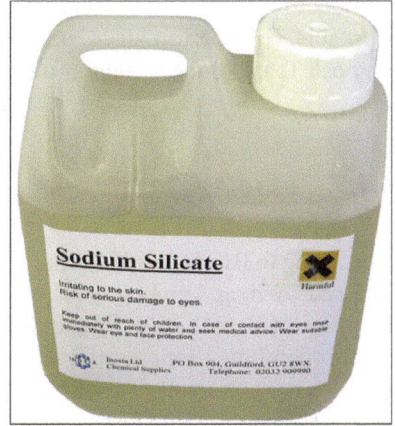

hydroxide or sodium hydroxide is used in geopolymer concrete as discussed by Goyal (2018). In geopolymer concrete, sodium silicate (Na$_2$SiO$_3$) is combined with sodium hydroxide (NaOH) to create an alkaline solution, as shown in Figs. 6.4 and 6.5. The sodium silicate enhances and speeds up the polymerization process.

For instance, if 12M NaOH is chosen, its molecular weight of 40 is multiplied by 12 to provide 480 g of solution. This indicates that 480 g of NaOH pellets have been dissolved in a liter of water. The solution is combined with 1.5–3 times the quantity of prepared NaOH solution of sodium silicate and the mixture is allowed to stand for 24 h. For the sodium silicate in this investigation, two ratios of 1.5 and 2.5 will be used with molarities ranging from 6 to 18 M.

The significant components of concrete are coarse and fine aggregates, which are combined with a binding glue to create concrete. This material improves the physical qualities of concrete because it exhibits excellent volume and stability for the mixture and resistance to various erosions. When used in concrete, they are divided into two categories: coarse aggregate and fine aggregate. While coarse aggregates are rough and large in texture and are more than 4.75 mm, they often consist of shattered stone pieces and gravel. On the other hand, fine aggregates are composed of fine particles, typically crushed stones, and sands, and may pass through a 4.75 mm sieve. It is an essential ingredient of concrete, with an aggregate content of between 60 and 80%. While the coarse aggregates are utilized as an inert filler for concrete mix, the fine aggregates assist in filling any gaps that may be present between the coarse aggregates' particles Mahmud (2021).

6.1.6 Super Plasticizer

To increase workability, a superplasticizer based on sulfated naphthalene formaldehyde is utilized.

According to IS 9103, a superplasticizer (SP) is employed to make concrete more workable. High performance sulfated naphthalene formaldehyde is employed for this purpose. 0.5% of the weight of the binding material is superplasticizer.

6.1.7 Fine Aggregate

According to BIS 383 (1970), fine aggregate that complied with Zone-2 is employed. The nearby river supplied the fine aggregate. The collected sand is sieved by using IS sieves of sizes 2.36, 1.18, 0.6, 0.3 and 0.15 mm. Sand that has been kept on each sieve is combined appropriately to create Zone-2 sand following the mix design, and the necessary amount of each size fraction is indicated in Table 6.2. Sand kept on each sieve was put into distinct bags and packed separately for usage; Table 6.3 lists their physical characteristics.

Table 6.2 Proportions of different size fractions of sand to obtain zone-II

Sieve size (mm)	Percentage passing recommended by IS 383 (1970)	Adopted grading	Percentage weight retained	Cumulative percentage weight retained	Weight retained in grams
10–4.75	100	100	–	–	–
4.75–2.36	90–100	100	–	–	–
2.36–1.18	75–100	90	10	10	100
1.18–0.60	55–90	65	25	35	250
0.60–0.30	35–59	40	25	60	250
0.30–0.15	8–30	10	30	90	300
0.15	0–10	0	10	100	100

Table 6.3 Physical properties of fine aggregate

Fineness modulus	2.59
Bulk density	1.45 gm/cc
Specific gravity	2.65

6.1.8 Coarse Aggregate

The coarse material utilized is crushed granite. The 20 mm nominal coarse material is supplied from a nearby crusher site. In this experiment, 20 mm well graded aggregate following IS 383 (1970) is employed. The coarse aggregate obtained from the quarry is sieved via sieves with corresponding mesh sizes of 80, 40, 20, 10 and 4.75 mm. The amount needed to mix each size fraction is provided in Table 6.4. The material kept on each filter is put in bags and stored individually and Table 6.5 displays the physical characteristics of the coarse aggregate.

6.2 Experimental Program

Before starting the research, numerous trial mixes of geopolymer paste, mortar and concrete specimens are cast to know the properties and evaluated for compressive strength.

The preliminary work's primary goals are:

1. To comprehend how fly ash and GGBS-based geopolymer concrete is made.

Table 6.4 Proportions of different size fractions of coarse aggregate

Sieve size (mm)	Percentage weight retained	Cumulative percentage weight retained	Percentage weight passing	Percentage weight passing for a graded aggregate of nominal size 20 mm (IS 383 1970)
80	0	0	100	–
40	0	0	100	100
20	0	0	100	95–100
10	70	70	30	25–55
4.75	30	100	0	0–10

Table 6.5 Physical properties of coarse aggregate

Fineness modulus	7.3
Bulk density	1.5 gm/cc
Specific gravity	2.80

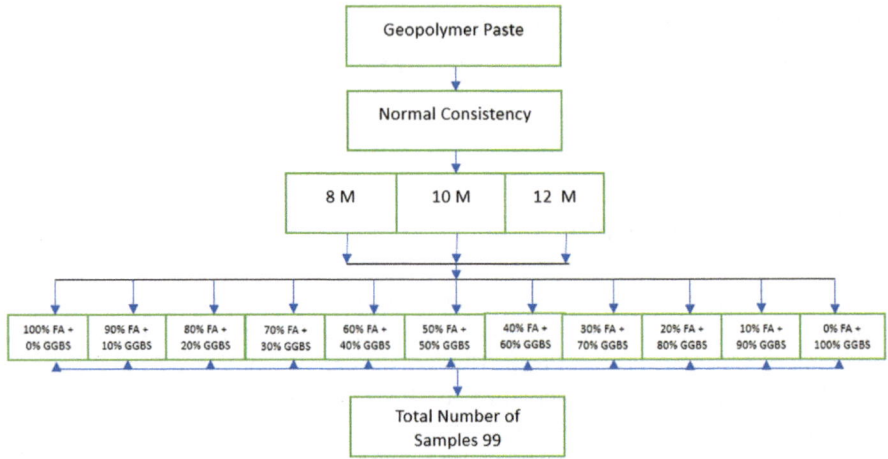

Fig. 6.6 Specimens cast to determine normal consistency of geopolymer paste

2. Knowing the impact of the sequence in which an alkaline activator is added to the ingredients in the mixer and observing the behavior of freshly mixed fly ash and GGBS-based geopolymer concrete are two objectives.
3. To create a mixing procedure and implement a curing schedule.
4. To comprehend the fundamental mix proportioning of geopolymer concrete based on fly ash and GGBS.
5. The following factors are considered in this study.
6. The total amount of binder.
7. The age of curing (7 and 28 days).
8. The curing regime is ambient and oven curing at 60 °C for 24 h.
9. The proportions of fly ash and GGBS and the ratio of alkaline to the binder.

6.2.1   Normal Consistency

To determine the Normal consistency of the source material, 99 samples are made with 10, 20, 30, 40, 50, 60, 70, 80, 90, and 100% substitution of GGBS in fly ash for 8M, 10M, and 12M. Figure 6.6 depicts the complete technique.

6.2.2   Final Setting Time

To assess the setting behavior of the source material, 99 samples are made with 10, 20, 30, 40, 50, 60, 70, 80, 90, and 100% substitution of GGBS in fly ash for 8M, 10M, and 12M. Figure 6.7 depicts the complete technique.

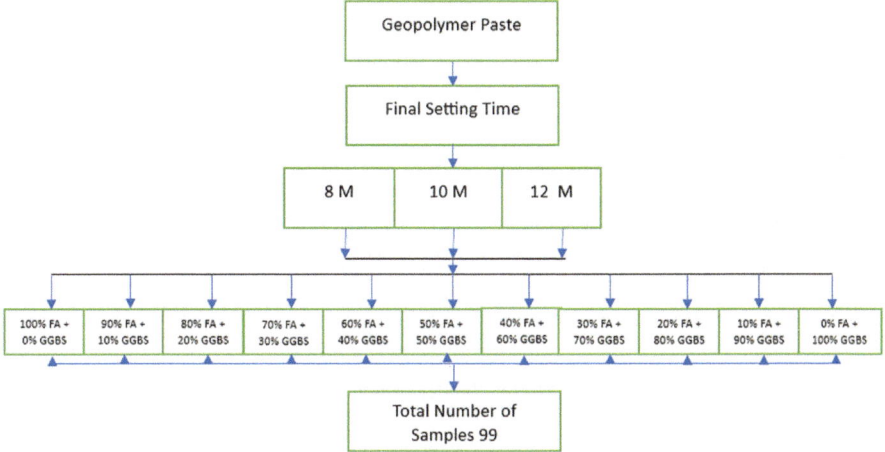

Fig. 6.7 Specimens cast to determine the final setting time of geopolymer paste

6.3 Compressive Strength of Mortar

To measure the compressive strength of mortar, 198 samples were made with varying substitutions of GGBS with fly ash for different amounts of sodium hydroxide (Fig. 6.8). Figure 6.9 depicts the comprehensive technique, while Table 6.8 shows the mix proportion details.

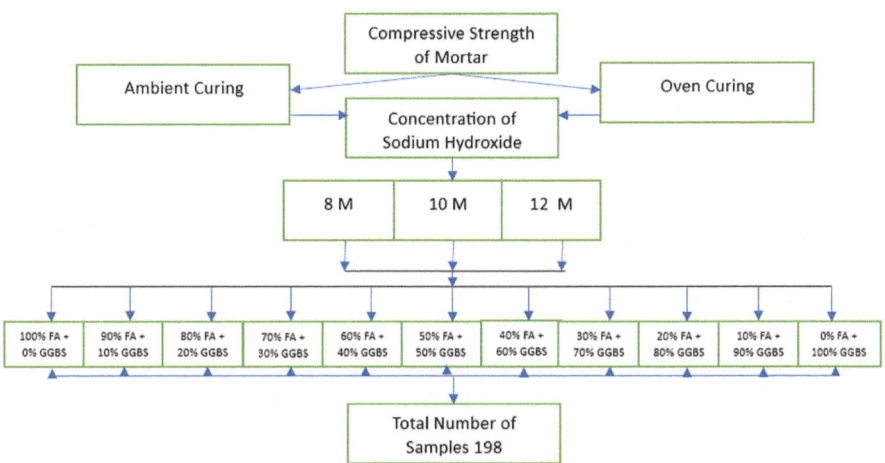

Fig. 6.8 Specimens cast to determine the compressive strength of mortar

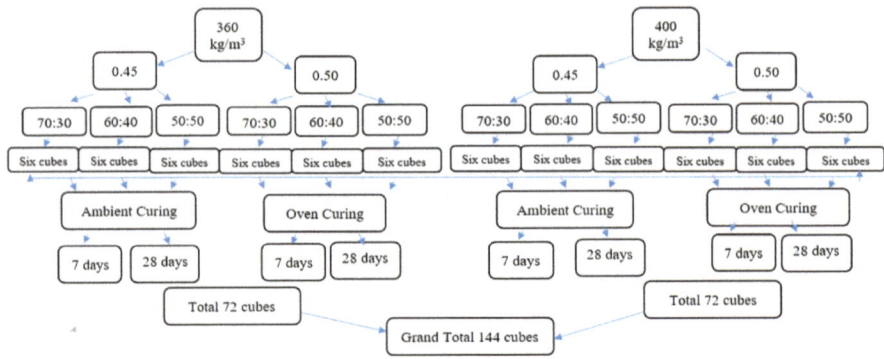

Fig. 6.9 Experimental program for different binder content

6.4 Quantities of Materials for Geopolymer Mortar for One Cubic Meter

The density of Mortar—2200 kg/m^3.

Volume $= 1$ m^3.

The strength ratio is maintained at 2.5.

Mix proportion $= 1{:}1$ is considered in this investigation.

Therefore, Weight of Binder Material $= (2200 \times 1)/(1 + 1 + 0.5) = 880$ kg/m^3 for one cubic meter.

Fine Aggregate $= 880$ kg/m^3 for one cubic meter.

Alkaline liquid $= 0.5 = 880/0.5 = 440$ kg/m^3. As mentioned above, the strength ratio (SS/SH $= 2.5$). Therefore, [Na$_2$SiO$_3$/ NaOH] $= 2.5$.

Hence, NaOH $= 125.72$ kg/m^3 and Na$_2$SiO$_3 = 314.29$ kg/m^3 for one cubic meter.

NaOH $+$ Na$_2$SiO$_3 = 125.72 + 314.29 = 440$ kg/m^3.

[(NaOH $(125.71) +$ Na2$_s$iO$_3$ $(314.29) = 440$].

6.5 Detailed Program of Compressive Strength of Concrete of Various Binder Content

Fly ash and GGBS, as well as the coarse and fine aggregate used in cement concrete, can be employed as binders to create geopolymer concrete Wang et al. (1995). Fly ash is abundant in alumina and silica and can generate aluminosilicate hydrate when it reacts with alkaline activators. The aluminosilicate hydrate, which forms due to the polymeric chain process, gives fly ash-based geopolymer concrete its strength. However, in the case of GGBS-based geopolymer concrete, the strength increase is caused by the gelization of calcium silicate during the polymerization with an alkaline solution. The qualities of geopolymer concrete based on GGBS are therefore superior to those of geopolymer concrete based on fly ash Roy (1999). Therefore, it can be claimed that the amounts of fly ash and GGBS, as well as the manner of curing, affect the strength and mechanical characteristics of geopolymer concrete. For the preparation of GPC, alkaline solutions, including sodium hydroxide (NaOH) and sodium silicate Na_2SiO_3 work well as alkaline activators. The fly ash and GGBS content ratios, the NaOH solution's molarity, the Na_2SiO_3/NaOH ratio, and the curing temperature are all variables that can alter and impact the compressive strength of concrete. According to research tests, a NaOH solution at 8–10 M concentration and a Na_2SiO_3/NaOH ratio of 2.5 produced the best compressive strength Hardjito et al. (2004). Getting enough strength requires curing as well. According to many research studies, fly ash-based GPC specimens' oven cured at 60 °C (24 h) had greater compressive strength than specimens cured outdoors at 25 °C Mustafa Al Bakri (2012).

Fly ash's silicon interacts with an alkaline solution during the polymerization process to create cementitious material. The poor reactivity of fly ash, which causes a long setting time and a slow rate of strength development, is the limiting factor for employing geopolymer concrete based on fly ash. It is noticed that substituting GGBS for fly ash helped improve compressive strength at young ages and eliminate oven curing. For a reaction to occur, fly ash-based geopolymers require an external energy source. In contrast, GGBS-based geopolymers eliminate external energy sources and achieve adequate strength during ambient curing, which might be advantageous in in-situ settings Gunneswara Rao et al. (2015).

6.6 Impact on the Sodium Silicate to Sodium Hydroxide Ratio

The relationship between the alkaline activators (NaOH and Na_2SiO_3) used in the manufacture of fly ash and GGBS-based geopolymer and their effects on the mechanical characteristics of GPC is investigated. For an experimental examination, geopolymer samples with various fly ash, GGBS, and Na_2SiO_3/NaOH ratios are created (1.5, 2.0, 2.5, and 3.0). The ratio and the curing conditions (ambient curing and oven curing at 60 °C for 24 h), the quantities of fly ash and GGBS (70–30 and 50–50), and the two grades of concrete are the characteristics taken into consideration in this research (GPC25 and GPC50) a similar study made by the author Gunneswara Rao et al. (2015) for M20 and M50 grade of concrete.

Tables 6.6 and 6.7 display the compressive strengths of geopolymer concrete for various Na_2SiO_3/NaOH ratios.

At a ratio of 2.5, the greatest compressive strength for M25 is measured at 40.18 MPa, while the highest compressive strength for M50 is 62.22 MPa. The increased sodium concentration, essential for polymerization, has enhanced compressive strength. The high OH- content in the mixes may have contributed to the poor

Table 6.6 Compressive strength of geopolymer of GPC25 (ambient and oven dried)

Na_2SiO_3/NaOH	7 days (N/mm^2)		28 days (N/mm^2)	
1.5	26.8	28.2	32.65	34.85
2	28.5	31.55	32.99	35.01
2.5	30.4	33.64	31.22	39.19
3	27.3	28.19	28.45	31.75

Table 6.7 Compressive strength of geopolymer of GPC50 (ambient and oven dried)

Na_2SiO_3/NaOH	7 days (N/mm^2)		28 days (N/mm^2)	
1.5	46.29	48.54	52.16	55.25
2	49.93	52.37	54.60	58.93
2.5	53.83	56.39	58.36	62.11
3	43.15	44.97	48.48	54.02

Table 6.8 Mix proportions of geopolymer concrete

Mix	Alkaline/ binder ratio	Binder content (kg/m^3) of concrete	Fine aggregate (kg/m^3) of concrete	Coarse aggregate (kg/ m^3)	Alkaline activator (kg/ m^3)
A2F70G30	0.45	360	774	1090.8	162
A2F70G30	0.5	360	774	1090.8	180
B1F70G30	0.45	400	810.6	966	189
B2F70G30	0.5	400	810.6	966	210

compressive strength of the geopolymer samples that are reported at ratio 3 (Xu and Van Deventer 2000).

The detailed practical program is shown in Fig. 6.9 (Table 6.8).

6.7 Casting and Curing of GPC

Separately weighing and mixing the dry ingredients takes place in a 100 kg rotating drum pan mixer. The alkaline liquid and superplasticizer are added after thoroughly combining the dry ingredients. Consequently, Fig. 6.10 displays a few examples of geopolymer concrete cubes. However, continued mixing for 5–7 min should provide homogenous mixing and ensure the viability of GPC. The table vibration procedure is employed for 45 s after the newly mixed concrete has been transferred into concrete molds (150 mm × 150 mm × 150 mm), and then the concrete is allowed to cure for 24 h. The cast specimens are properly demolded and cured after 24 h. Until the required testing age, cured specimens are placed in the open air at 25 °C and 65%

Fig. 6.10 Few samples of geopolymer concrete cubes

relative humidity for 7 and 28 days. The demolded samples are held at 60 °C for 24 h as part of the oven curing procedure. After that, they are removed from the oven and maintained at room temperature for a predetermined time (7 or 28 days).

6.8 Durability Properties of Geopolymer Concrete

To assess the effect of elevated temperatures on the fly ash and GGBS-based GPC. Three different temperatures are exposed to 200°, 400° and 600 °C and every specimen for 30, 60 and 90 min. Air cooling and water quenching processes are used for M25 grade, fly ash: GGBS-70:30, A/B ratio is 0.55, and for M50 grade, fly ash: GGBS is 50:50, Alkaline/Binder ratio is 0.50. The details of the experimental program and mix proportion details are provided in Tables 6.9 and 6.10 and the detailed experimental program is shown in Fig. 6.11.

Table 6.9 Details of specimens cast for elevated temperatures

Grade of concrete	Temperature (°C)	Number of samples for cast (air cooling)			Number of instances of cast (water quenching)		
		30 min	60 min	90 min	30 min	60 min	90 min
M25	200	3	3	3	3	3	3
	400	3	3	3	3	3	3
	600	3	3	3	3	3	3
M50	200	3	3	3	3	3	3
	400	3	3	3	3	3	3
	600	3	3	3	3	3	3
Total no of specimens cast		54			54		

Table 6.10 Quantities of geopolymer concrete for M25 and M50 grades for one cubic meter

Mix	Fly ash (kg/m^3)	GGBS (kg/m^3)	Fine aggregate (kg/m^3)	Coarse aggregate (kg/m^3)	Na$_2$SiO$_3$ (kg/m^3)	NaOH (kg/m^3)
M25	252	108	770.05	1090.8	141.42	56.57
M50	225	225	761.02	973.00	160.71	64.28

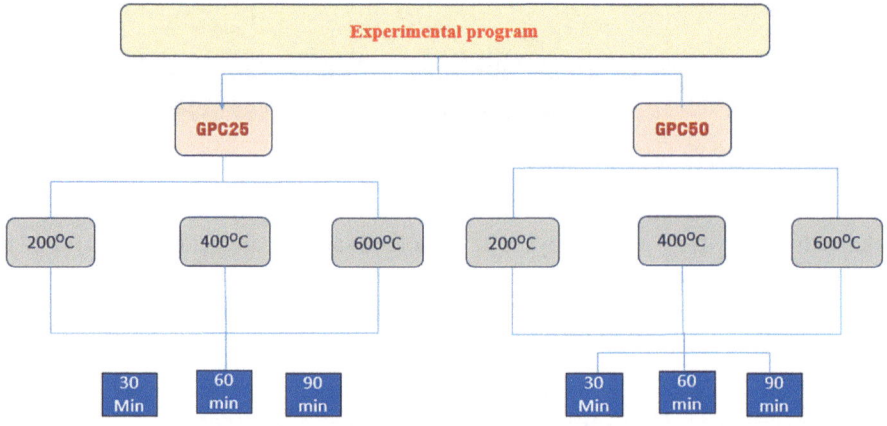

Fig. 6.11 Experimental program for different grade of concrete

6.9   Geopolymer Concrete Mix Design Approach

Using the Indian standard mix design stages of cement concrete and a trial-and-error method, a straightforward mix design technique is assessed for fly ash and GGBS-based GPC.

Step 1: Target Strength (ft) Calculation: fck + 1.65 SD

ft = The desired geopolymer concrete's average compressive strength after 28 days.

fck stands for the geopolymer concrete's typical compressive strength after 28 days.

SD = Standard Deviation.

Step 2: Selecting a Slump

The kind of construction project will determine the slump, and the correct figure may be guessed based on the workability requirement.

Step 3: Choosing the Alkaline/Binder Ratio and the Binder Proportion

Compressive strength and workability determine binder percentage. When strength needs to rise, fly ash must be replaced with GGBS. Workability also varies with binder quantity. High fly ash in GPC increases workability but not compressive strength, while high GGBS decreases workability but increases compressive strength. The ratio of alkaline solution of NaOH and Na_2SiO_3 to binder of fly ash and GGBS in a geopolymer concrete mix affects concrete quality and compressive strength. A stiffer, lower-compressive-strength combination has a lower alkaline/binder ratio. The alkaline/binder ratio affects compressive strength. For example, the mixture is stiff at 0.45 alkaline/binder and separated at 0.5. Changing binder contents, aggregate-binder ratios, aggregate size and other characteristics can provide varying compressive strengths for ambient and oven curing. Outdoor or oven curing may be used depending on the building.

Step 4: Choosing the Aggregate/Binder Ratio

During ambient and oven curing, the projected compressive strength for fly ash and GGBS-based geopolymer concrete at various alkaline-binder ratios. According to findings, compressive strength falls as the aggregate-binder ratio rises, and this drop is comparable to that of regular concrete. The selection of the aggregate-binder ratio is influenced by the required strength, binder percentage and alkaline-binder ratio. The difference in compressive strength between GGBS with a variable binder percentage and geopolymer concrete with flyash is just a few percent. As a result, it doesn't seem that aggregate content affects the strength of high strength geopolymer concrete.

Step 5: Choosing the Content of the Binder to Achieve the Necessary Target Strength

In the experiment, two binder components with fly ash and GGBS proportions are used. Although the findings for compressive strength are based on these three binder contents, linear interpolation may be used to determine compressive strength for intermediate binder concentrations. The binder content must be selected to reach the requisite strength at the highest binder concentration. The amount of fly ash and GGBS, as well as the alkaline-binder ratio, also have a role. Compared to fly ash doses, the amount of binder needed for higher degrees of GGBS is low for the same compressive strength. The amount of binder required depends on the curing temperature as well. The choice of binder concentration relies on factors such as strength, the accessibility of the raw materials, workability, curing temperature, etc.

Step 6: Estimation of the Fine and Coarse Aggregate Content

Figure 6.12 depicts the fluctuation in binder content and the coarse aggregate-to-total aggregate ratio. The figure shows that the coarse aggregate total amount is 0.570 for 400 kg/m^3 binder content. The coarse aggregate-total aggregate for specific binder contents may be approximated through these numbers. The acquired numbers may determine the coarse and fine aggregate amounts.

$$\text{Fine aggregate} = \text{Total aggregate} - \text{Coarse aggregate}$$

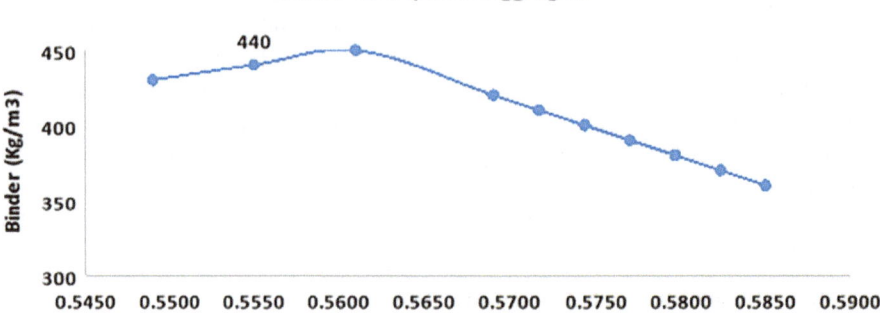

Fig. 6.12 Variation of coarse aggregate-total aggregate ratio w.r.t. binder content (360–400 kg/m^3), Gunneswara Rao et al. (2015)

Step 7: Finding Alkaline Content

The process for determining the alkaline-binder ratio, binder percentage and binder contents is described in the preceding sections. If binder, alkaline solution and amounts are obtained for 1 m^3 of concrete, the residual coarse aggregate and fine aggregate contents may be calculated by subtracting the acquired amounts. According to literature and experiments, the ideal mass ratio of sodium hydroxide solution to sodium silicate solution is 2.5. Thus, the amounts of NaOH and Na$_2$SiO$_3$ are derived from the chosen alkaline-binder ratio and the molarity of NaOH is chosen as 8M.

6.10   Example of a Mix Ratio Following the Suggested Method

Here, a GPC mix design is shown as an illustration of the proposed process.

- Assume that the fly ash and GGBS-based geopolymer concrete can be made to achieve the desired strength of 25 MPa for a 120 mm slump at 28 days. The strength requirement and workability indicate that the concrete is of low grade, where the strengths may be attained by using fewer amounts of binder and GGBS. The binder percentage of 70–30, and an anticipated binder concentration of 400 kg/m^3. The aggregate-binder ratio from the studies taken by Gunneswara Rao et al. (2015), the Aggregate/Binder content equals 4.54, is calculated by taking the alkaline-binder ratio to be 0.5. A similar method is deployed in this mixed design.

  The ratio of Na$_2$SiO$_3$/NaOH is 2.5, and an 8M NaOH Solution is considered. The compressive strength and workability of GPC will decrease when the alkaline-to-binder ratio decreases. Workability guidelines provided by IS 456 (2000) are

Table 6.11 Quantities in one cubic meter for M25 geopolymer concrete

Material	Quantity in kg
Binder	400
Fly ash: GGBS	70:30
Coarse aggregate	1035.12
Fine aggregate	780.88
Alkaline solution	200

broken down into Low, Medium, and High categories. The combination is chosen based on its durability and workability.

- The total alkaline content is calculated by multiplying the Alkaline/Binder ratio by the total fly ash quantity: GGBS.
- $0.5 \times 400 = 200$ kg in one cubic of concrete.
- By knowing the ratio of $Na_2SiO_3/NaOH = 2.5$, the individual quantities of NaOH and Na_2SiO_3 can be determined.
- NaOH solution $= 200/3.5 = 57.14$ kg in one cubic of concrete.
- Na_2SiO_3 Solution $= 2.5 \times 57.14 = 142.85$ kg in one cubic of concrete.
- Total Aggregate/Binder ratio $= 4.54$.
- Total Aggregate $= 4.54 \times 400 = 1816$ kg in one cubic of concrete.
- From the studies of Gunneswara Rao et al. (2015), the Coarse Aggregate/Total Aggregate for 400 kg/m^3 = 0.5700.
- Coarse Aggregate $= 0.5700 \times 1816 = 1035.12$ kg in one cubic of concrete.

Fine aggregate = Total aggregate − Coarse aggregate

-
$$= 1816 - 1035.12 = 780.88 \, kg/m^3$$
$$= 776.39 \, kg \text{ in one cubic of concrete.}$$

The mix proportions for geopolymer concrete with a 25 MPa compressive strength that is cured ambient and in an oven are shown in Table 6.11. Concrete cubes measuring $150 \times 150 \times 150$ mm are cast using the determined mix proportions and cured using the stated curing process.

6.11 Normal Consistency of Geopolymer Paste

Table 6.12 presents the typical consistency results for various amounts of replacement of the original content. According to the findings, there is an increase in average consistency values for mixes with an increase in GGBS concentration. This is because the GGBS particles have a sharp and flaky elongated morphology, resulting in substantially more internal friction than fly ash particles. As a result, a more alkaline solution is required to attain the requisite normal consistency. Similarly, 100%

fly ash requires a less alkaline activator to obtain the same 28% consistency as that manufactured with 100% GGBS because fly ash particles need less solution.

Furthermore, the normal consistency rises when the GGBS level increases. For example, the normal consistency rises from 28 to 33% when 70% fly ash and 30% GGBS are used. Therefore, the consistency of the alkaline solution is substantially higher than that of water, distinguishing it from cement paste. Hence, geopolymer concrete cannot be utilized in its green condition unless mixed with an alkaline solution. Furthermore, increasing the amount of sodium hydroxide does not affect the consistency. As a result, increasing GGBS leads to higher normal consistency.

Consistency

The fly ash containing geopolymer paste required less alkaline content to obtain better consistency than the GGBS containing paste. Increasing the GGBS content in intermediate mixes increased standard consistency. Fly ash exhibits less internal friction due to its spherical shape, so Vicat's plunger functions with a lower alkaline activator content. From the SEM analysis, the GGBS particles have elongated, straight, flaky and sharp angular surfaces which exhibit more crystalline structure. They have a higher level of internal friction than fly ash particles, which require a high amount of solution to achieve better consistency. When combined with fly ash at 80%, GGBS has a consistency of 37%.

Final Setting Time

Table 6.13 shows how variations in 8M, 10M and 12M sodium hydroxide concentration and GGBS proportions in fly ash are utilized to evaluate the setting behavior of geopolymer. These findings pertain to geopolymer paste ultimate setting times composed of various fly ash and GGBS. Determining the setting time of fly ash and GGBS paste (Binder) is the same as deciding the set time of cement. The mix's

Table 6.12 Results of normal consistency with 8M, 10M, and 12M

Binder material		The concentration of NaOH		
Fly ash	GGBS	8M	10M	12M
100	0	28	28	28
90	10	27	27	29
80	20	31	30	31
70	30	33	31	32
60	40	33	33	32
50	50	33	33	33
40	60	33	33	35
30	70	33	35	38
20	80	37	37	39
10	90	37	37	39
0	100	37	37	39

ultimate setting time changes as sodium hydroxide's molarity varies with fly ash and GGBS. Table 6.13 illustrates this. According to Wan et al. (2004), the alumina content (Fly ash and GGBS) had a substantial influence on the setting time of GPC. The setting time of the mix lowers as the alumina concentration increases. As a result, GGBS mixed with fly ash might be recommended to achieve the appropriate setting time. The normal consistency of 28% with 100% fly ash, used to estimate setting time is 0.85P (0.85 × 28 = 23.8%). Similarly, the alkaline activator used to compute the setting time behavior of 100% GGBS-based paste is 0.85P (0.85 × 37 = 31.45%).

Setting Time Behavior

Fly ash's reactive process is slower than GGBS, indicating that it is a more reactive compound. Because of its low reactivity, fly ash is slow to set and develop strength. It often does not dissolve entirely before hardening. During alkali activation of ash/slag blends, Mullite in fly ash remains unreacted and calcium appears active. GGBS is faster at setting when used with an alkaline activator than fly ash. Therefore, GGBS is a better source of raw material for geopolymer materials with high early strength than fly ash. When GGBS is replaced by 20%, the setting time decreases to 125 min from 200 min (for 8M sodium hydroxide). According to the test results, the final setting time has significantly reduced for the 8M mix when GGBS is used instead of total fly ash. Because of geopolymers' quick setting time, they cannot be used in conventional construction. Therefore, GGBS and fly ash are suitable for achieving better setting behavior.

Table 6.13 Results of final setting time of paste for 8M, 10M, and 12M

Binder material		Final setting time in minutes for different molarities		
Fly ash	GGBS	8M	10M	12M
100	0	180	200	220
90	10	170	185	190
80	20	165	175	180
70	30	150	155	170
60	40	120	130	125
50	50	90	105	100
40	60	80	85	90
30	70	70	76	90
20	80	55	60	70
10	90	42	48	60
0	100	40	42	50

Compressive Strength of Mortar

The raw materials fly ash + GGBS and alkali binder compressive strength ratios are 1:1. In this experiment, the mixture of both is kept constant. Table 6.14 detail the mortar results with various sodium hydroxide concentrations.

The Compressive Strength of Mortar

Various compressive strengths are produced by utilizing an alkaline activator with varying concentrations of sodium hydroxide and fly ash: GGBS ratios. Geopolymer mortars have compressive strengths ranging from 39 to 75 MPa. Changing the sodium hydroxide molarity resulted in compressive strengths ranging from 39 to 70 MPa for ambient cured materials. The compressive strength is increased by increasing the concentration of NaOH in the mortar. This might be due to the higher sodium hydroxide content in the alkaline activator. 8M, for example, comprises 320 g of NaOH, whereas 12M contains 480 g. They increase the quantity of fly ash substitution with GGBS enhanced mortar strength in ambient curing samples. According to a previous study, premature polymerization causes oven cured specimens to be stronger than ambient cured specimens. However, because oven curing is not possible at the site, my investigation will concentrate on the effect of GGBS and ambient curing on specimen strength. Finally, the results demonstrated that the activity of GGBS is crucial in obtaining oven curing strength. Because fly ash and GGBS have a close relationship between the aggregates and homogeneity of the geopolymer paste, the increase in strength may be due to the chemical reaction during polymerization. When

Table 6.14 Compressive strengths of geopolymer mortars in N/mm^2 for different concentrations of sodium hydroxide

Binding material		Compressive strength (N/mm^2) for various concentrations of sodium hydroxide					
Fly ash	GGBS	8M ambient curing	8M oven curing	10M ambient curing	10M oven curing	12M ambient curing	12M oven curing
0	100	39	41	41	42	42	44
10	90	42	43	43	44	44	47
20	80	44	45	45	47	47	51
30	70	45	46	47	48	48	53
40	60	47	48	48	49	49	55
50	50	48	50	50	52	52	60
60	40	50	51	52	55	55	64
70	30	52	53	54	57	57	66
80	20	54	59	59	60	60	68
90	10	55	62	63	65	65	70
100	0	59	64	66	70	70	75

the GGBS concentration increases, there is a slight difference in strength between ambient and oven curing, indicating that low molarity sodium is employed.

When blended with 50% fly ash and 50% GGBS, the strength is roughly 48 MPa. Similarly, the compressive strength of 10M and 12M under outdoor and oven curing conditions are 66% and 70%, and 70% and 75% respectively. When the GGBS concentration is raised from 0 to 50%, the strength improves by approximately 7 MPa. After oven and outdoor curing, Geopolymer mortar has compressive strengths of about 75 and 70 MPa for 100% GGBS content. Geopolymers can be as strong as concrete when GGBS and fly ash are blended as source materials. The increased calcium concentration of GGBS may account for its higher compressive strength of 32.6%. The geopolymer mortar based on fly ash surpasses the ambient sample. Oven curing allows for faster production of geopolymers and accelerates the polymerization process, resulting in faster-enhanced strength. The mortar matrix's porosity strengthens the interaction in providing crystalline structure. The alumina bonds are formed initially when polymerization begins. However, silicon bonds develop more and lead to polymerization, which makes geopolymers strong as concrete with GGBS and fly ash blended as source materials. The presence of more calcium in GGBS adds to improved compressive strength.

6.12 Fresh and Hardened Properties of Geopolymer Concrete

The Workability of Geopolymer Concrete

Geopolymer concrete is challenging to compact due to its rigid consistency when new. The only way is to make geopolymer concrete workable to utilize high range admixtures that reduce water content, particularly naphthalene-based superplasticizers. A binder dosage of 4% by mass is required to complete this trial. The slump values of geopolymer concrete are determined using a slump test, as indicated in Table 6.15. Based on slump values, the slump is visible at high alkaline-binder ratios, binder concentrations and fly ash levels. This might be due to higher binder content; therefore, more fly ash particles are used. Since fly ash particles are spherical, the alkaline-binder ratio increased and decreased slump values with the same binder content. This is also due to concrete exhibiting a more crystalline structure.

Due to the lower alkaline concentration, an additional superplasticizer is necessary to achieve optimum workability because GGBS particles are angular and react quickly. Substituting GGBS for fly ash reduced slump values.

Table 6.16 shows that the compressive strength improves for all binder compositions as GGBS content increases. Due to the production of a rich calcium silicate hydrate gel, the compressive strength has risen when fly ash has been replaced with GGBS. Fly ash is gradually replaced with GGBS, which improves strength while speeding up the setting. Results show that the compressive strength is highest with an

Table 6.15 Workability of geopolymer concrete

Fly ash: GGBS	Alkali-binder ratio	Binder quantity (kg/m^3)	
		360	400
70:30	0.45	95	102
	0.50	102	125
60:40	0.45	84	100
	0.50	100	110
50:50	0.45	78	92
	0.50	92	101

alkaline/binder ratio of 0.5. The same outcomes are obtained with oven-cured specimens as well. It is clear from the results that oven-cured samples exhibited higher compressive strength when fly ash is replaced at a lower level by 30%. The compressive strength values of the oven-cured and outdoor-cured specimens are practically identical when the replacement level is raised by 40% and 50%. This suggests that by substituting GGBS at the proper percentages for fly ash, it is possible to increase the applicability of GPC under field circumstances.

Compressive strength and slump values for intermediate binder contents at specific curing temperatures, curing period, and NaOH with 8M concentration can be estimated using existing graphs by considering aggregate/binder, alkaline/binder, fly ash and GGBS proportion, slump value, and compressive strength as variables (8M).

Table 6.16 Compressive strength of geopolymer concrete with 360 and 400 kg/m^3 binder quantity

Binder content (kg/m^3)	Source material (fly ash: GGBS)	A/B ratio	Type of curing and strength			
			Ambient curing		Oven curing	
			7 days (N/mm^2)	28 days (N/mm^2)	7 days (N/mm^2)	28 days (N/mm^2)
360	70:30	0.45	17.28	32.26	18.25	35.25
		0.50	18.45	33.25	19.35	37.25
	60:40	0.45	19.20	34.28	20.50	38.10
		0.50	19.45	34.30	21.50	39.15
	50:50	0.45	21.56	35.35	23.56	39.20
		0.50	21.50	36.25	23.65	40.25
400	70:30	0.45	27.20	39.16	28.55	41.25
		0.50	28.10	40.15	29.68	42.20
	60:40	0.45	29.35	41.10	30.15	42.30
		0.50	29.45	42.45	31.10	43.30
	50:50	0.45	31.56	44.50	33.20	45.40
		0.50	31.45	45.60	33.25	47.35

It is crucial to research how the components of the geopolymer concrete mix affect its strength over time. To produce concrete with the desired goal strength, a mix design is a particular mixture of raw ingredients. Therefore, the composition of the concrete plays a significant role in achieving 28 days compressive strength. It is well known that raising the concrete's alkaline component enhances its strength. In addition, the quantity of fine and coarse aggregate used in the mix and the binder concentration affect the concrete's strength. A mixed design technique for fly ash and GGBS-based GPC is suggested based on the findings of the predicted compressive strength calculations. This methodology may enable the design engineers to create GPC at any desired compressive strength.

The Effect of Curing on the Compressive Strength of Concrete

This study investigates ambient and oven curing impacts on compressive strength. Geopolymer has a higher compressive strength in the range than it does ambient. The polymerization process is speed up at a higher temperature than the ambient temperature, increasing the strength. Therefore, concrete should be cured at room temperature in many practical applications. When the molarity of sodium hydroxide is 8M, the GPC specimens cured at outdoor temperatures reached a maximum strength of roughly 45 MPa for 400 kg/m^3, whereas 47 MPa at oven temperature. The result shows a similar value with a slight increase of around 0.5% higher in oven curing compared to outdoor curing. The Outdoor curing of fly ash and GGBS-based geopolymer concrete is possible even with a low NaOH molarity. where oven curing is difficult in the field. As a result, GPC formed from GGBS, which replaces fly ash, can be manufactured without an oven. The compressive strength values obtained from the experimental work are shown in the results for both 7 and 28 days. As the proportion of GGBS in the mix is altered with increasing alkaline content, the compressive strength increased after seven days and lasted for 28 days. After 28 days, GGBS-based mixtures with 50% GGBS content outperformed GGBS-based geopolymer concrete. The enhanced calcium concentration of GGBS contributed to the combination's increased strength. The inclusion of soluble calcium increases compressive strength and speeds up the hardening process.

The Strength of Concrete as a Function of Age

Even though concrete strengthens after 28 days, it is frequently referred to as 28 days concrete strength. GPC's 7- and 28 days strengths have a significant correlation. The molarity of an alkaline activator in geopolymer concrete is proportional to the binder concentration and curing regime. Strength develops faster at a young age than at an older age. For both types of curing, the strength gain is more prominent for oven-cured GPCs than for outdoor-cured GPCs. After seven days of curing, the compressive strength of the oven cured specimen is greater than that of the outdoor cured sample. The initial rate of strength gain is considerable, but it does not last as long as regular concrete. The ratio of 28 days compressive strength to 7 days compressive strength demonstrates this.

The 28 days old ambient and oven cured samples are shown in Table 6.17. For both ambient and oven curing, the 28 days test results for splitting tensile strengths for 360 and 400 kg/m^3 range from 1.87 to 3.25 MPa. The higher value indicated for the 400 kg/m^3 and 0.5 alkaline binder ratio with around 16% higher in oven curing compared to outdoor curing. This demonstrates that ambient curing allows GPC specimens to acquire their split tensile strength. This is apparent due to the geopolymer gel's steadfast attachment to the aggregate particle. With the addition of GGBS to the fly ash at 28 days, the split tensile strength of GPC specimens cured ambient obtained enough strength, and with the rise in GGBS, there is also tensile strength. Aluminosilicate hydrate and calcium silicate hydrate geopolymeric gels continue to develop in geopolymer concrete, and both fly ash and GGBS contribute to this process, as stated by Nath et al. (2014). The relationship between compressive and tensile strength in ordinary concrete is $0.7\sqrt{fck}$. However, in this case, the number is lower in the studies of Gunneswara Rao et al. (2015).

Table 6.18 displays the flexural strength of geopolymer concrete at 28 days. Due to the quick polymerization process, the strength of oven cured concrete specimens is greater than that of ambient cured specimens. The flexural strength of oven cured specimens of geopolymer concrete is around 3.28 MPa, while it is about 2.96 MPa for 400 kg/m^3 when it is ambient cured.

The flexural strength of the concrete increases with an increase in GGBS concentration. The GGBS content is crucial to the mixture's ability to produce CASH gel. This might be the mortar matrix's porosity and strengthening interaction with the aggregate; hence, concrete exhibits a more crystalline structure. Therefore, the strength contribution is due to this gel formation. While the formation of sodium aluminosilicate gel serves as the primary basis for the compressive strengths of fly ash-based geopolymer concretes, in the case of geopolymer with GGBS, some

Table 6.17 Split tensile strength for geopolymer concrete

Binder content (kg/m^3)	Binding material	Alkaline/binder ratio	Split tensile strength after 28 days (N/mm^2)	
	Fly ash: GGBS		Ambient curing	Oven curing
360	70:30	0.45	2.05	2.7
		0.50	1.84	2.95
	60:40	0.45	1.87	2.58
		0.50	1.92	2.74
	50:50	0.45	1.98	2.95
		0.50	2.3	2.83
400	70:30	0.45	2.36	2.76
		0.50	2.21	2.85
	60:40	0.45	2.52	2.83
		0.50	2.45	3.25
	50:50	0.45	2.39	2.94
		0.50	2.45	3.00

Table 6.18 Flexural strength of geopolymer concrete

Binder content	Binding material	Alkaline/binder ratio	Flexural strength after 28 days (N/mm^2)	
	Fly ash: GGBS		Ambient curing	Oven curing
360	70:30	0.45	2.49	2.47
		0.50	2.4	2.55
	60:40	0.45	1.95	2.67
		0.50	2.28	2.56
	50:50	0.45	2.49	2.68
		0.50	2.4	2.76
400	70:30	0.45	2.8	2.8
		0.50	2.56	2.74
	60:40	0.45	3.04	2.92
		0.50	2.96	3.28
	50:50	0.45	2.4	2.6
		0.50	2.8	2.93

calcium bearing binder compounds are accountable for the strength contribution in an increase in the flexural strength.

6.13   Conclusions

The current study examines the impact of GGBS on the strength and durability characteristics of geopolymers and mineralogical research. The results of the study's conclusions are displayed below.

6.13.1   Normal Consistency, Final Setting Time, and Compressive Strength of Geopolymer Mortar

1. The normal consistency of the geopolymer is not significantly impacted by the molarity of sodium hydroxide in the alkaline activator.
2. As the sodium hydroxide solution concentration rises, so does the ultimate setting time.
3. The final setting time of the geopolymer paste is lowered by substituting GGBS for fly ash.
4. All combinations of fly ash and GGBS have greater compressive strengths when the sodium hydroxide solution content is increased.

5. The proportion of GGBS in the mix enhances the compressive strength of geopolymer mortar.
6. One potential method for creating geopolymer under ambient curing circumstances is the combination of fly ash and GGBS.
7. The compressive strength of GGBS and fly ash increases as the sodium hydroxide solution concentration rises.
8. Even if the alkaline activator includes sodium hydroxide, the sodium hydroxide does not substantially alter the behavior of geopolymer mortar.
9. Geopolymer mortar that contains more GGBS has greater strength.
10. The curing procedure is crucial to the polymerization process.

6.13.2   Workability and Mechanical Properties of Geopolymer Concrete

1. It is demonstrated that GGBS, an industrial waste product, can be used to manufacture geopolymer concrete.
2. The compressive strength of fly ash and GGBS-based geopolymer concrete increases as the concentration of sodium hydroxide solution rises.
3. Geopolymer mortar is more durable when the GGBS component is greater.
4. GGBS mixed with fly ash can aid in the creation of geopolymer mortar in outdoor curing conditions.
5. The curing process is crucial for polymerization.
6. Extensive testing on fly ash and GGBS-based geopolymer concrete reveals that substituting GGBS for fly ash enhances concrete compressive strength, independent of curing. The advantage is more pronounced for outdoor curing.
7. The critical temperature for geopolymer concrete based on fly ash and GGBS has been determined to be 60 °C.
8. The effect of GGBS on workability is inversely proportional to the strength of concrete.
9. The compressive strength of GGBS increased when 50% of the GGBS was replaced with fly ash for a binder content of 360 kg/m^3, and a similar tendency was noticed at binder amounts of 400 kg/m^3.
10. The proposed methodology is used to conduct workability and compressive strength tests on mixes, yielding reliable workability and compressive strength results.
11. Under ambient curing circumstances, combining fly ash and GGBS could be a viable solution for making geopolymer concrete.
12. When GGBS is added, geopolymer concrete sets much faster and provides greater strength without oven curing.

6.13.3 Durability Properties of Geopolymer Concrete

1. A higher temperature and a longer exposure duration resulted in an increased loss of GPC mass after 600 °C. As a result, there is a noticeable loss of mass.
2. Due to its exposure to heat for 90 min at a temperature of 200 °C, GPC 25 showed increased compressive strength and reduction at 600 °C.
3. For GPC50, the compressive strength is significantly reduced at 200 °C for 30 min.
4. With increased temperatures and extended exposure periods, XRD analysis revealed that geopolymer concrete exhibits a more crystalline structure.
5. When GPC is cooled from a high temperature, it significantly decreases the compressive strength for M25 at 600 °C compared to water quenching.
6. Since GPC 50 contains more calcium than GPC 25, the matrix produces more gypsum. The declassification of the aluminosilicate bond (CASH) gel and the rise in gypsum content are responsible for this strength loss caused by acid exposure.
7. It has been shown that the Acid Mass Loss Factor in the presence of sulfuric acid is smaller in the case of GPC 25 than in GPC 50.
8. Strength deterioration increases with higher GGBS content. The leaching and dissolving of the binder and the breakdown of degradation products inside the structure are two possible causes of the alkali-activated material's deterioration in an acidic solution.

6.13.4 Area of Future Study

1. High-strength concrete with lower molarity must be developed for geopolymer concrete during outdoor curing to achieve the standard grade.
2. The shrinkage and creep properties of fly ash and GGBS-based materials are crucial to understand.
3. Researching the structural characteristics of concrete made with fly ash and GGBS.

References

K. Basham, M. Clark, T. France, P. Harrison, *What is Fly Ash?* [Online] (Materials—Concrete Construction Magazine 2007)

BIS (Bureau of Indian Standards), *IS 383-1970: Specification for Coarse and Fine Aggregates from Natural Sources for Concrete* (Bureau of Indian Standard, New Delhi, India, 1970)

Boral Resources (2020) *What is the Difference Between Class F and Class C? Fly Ash* [Online]. Available from https://flyash.com/reference/, 27 Mar 2021

CSMA (2021) Cementitious Slag Makers Association. What is GGBS?—CSMA-The Cementitious Slag Makers Association (ukcsma.co.uk)

P.S. Deb, P. Nath, P.K. Sarker, The effects of ground-granulated blast-furnace slag blending with fly ash and activator content on the workability and strength properties of geopolymer concrete cured at ambient temperature. Mater. Des. **62**, 32–39 (2014)

S. Goyal, *What is an Alkali and Alkaline solution?* [Online]. jagranjosh.com (2018)

T. D. Gunneswara Rao, P. Alfrite, G. Mallikarjuna Rao, M. Andal, Fracture parameters of fly ash and GGBS-based geopolymer concrete. Appl. Mech. Mater. **764–765**, 1090–1094 (2015)

D. Hardjito, S.E. Wallah, D.M. Sumajouw, B. Rangan, Factors influencing the compressive strength of fly ash-based geopolymer concrete. Civ. Eng. Dimen. **6**(2), 88–93 (2004)

P. Hemali, *Fly Ash for Concrete: Properties, Uses, Advantages & Disadvantages.* gharpedia.com (2019)

IS 456, *Plain and Reinforced Concrete-Code of Practice* (BIS, New Delhi, 2000)

A.M. Mustafa Al Bakri, H. Kamarudin, M. Bnhussain, A.R. Rafiza, Y. Zarina, Effect of Na_2SiO_3/ NaOH ratios and NaOH molarities on compressive strength of fly-ash-based geopolymer. ACI Mater. J. **109**(5) (2012)

D.M. Roy, Alkali-activated cements opportunities and challenges. Cem. Concr. Res. **29**(2), 249–254 (1999)

H. Wan, Z. Shui, Z. Lin, Analysis of geometric characteristics of GGBS particles and their influences on cement properties. Cem. Concr. Res. **34**(1), 133–137 (2004)

S.D. Wang, X.C. Pu, K.L. Scrivener, P.L. Pratt, Alkali-activated slag cement and concrete: a review of properties and problems. Adv. Cem. Res. **7**, 93–102 (1995)

H. Xu, J.S.J. Van Deventer, The geopolymerisation of alumino-silicate minerals. Int. J. Miner. Process. **59**(3), 247–266 (2000)

Chapter 7
Case Investigations of Geopolymer Concrete

This chapter discusses the uses in different construction projects and their performance in different environmental conditions.

Geopolymer concrete has been successfully implemented in numerous construction projects globally, demonstrating its versatility and efficacy across diverse environmental conditions. This innovative material has garnered significant attention in the construction industry due to its superior performance characteristics and reduced environmental impact compared to traditional Portland cement concrete.

A notable case study illustrating the practical application of geopolymer concrete is the Brisbane West Well Camp Airport in Australia, completed in 2014. This project utilized geopolymer concrete for its taxiway, apron, and heavy-duty pavement areas, encompassing a substantial portion of the airport's infrastructure. The implementation of geopolymer concrete in this project demonstrated the material's capacity to withstand heavy loads from aircraft and resist chemical attacks from fuel spills and other potentially corrosive substances. Furthermore, the use of geopolymer concrete in this project resulted in a significant reduction in carbon emissions, with approximately 6600 tons less CO_2 produced compared to the use of traditional concrete. This substantial decrease in environmental impact underscores one of the key advantages of geopolymer concrete in large-scale construction projects.

Another significant application of geopolymer concrete was in the construction of the Global Change Institute building at the University of Queensland. In this project, geopolymer concrete was utilized for various structural elements, showcasing its versatility in building construction. The implementation of geopolymer concrete in this building resulted in a remarkable 70% reduction in carbon emissions compared to conventional concrete. This case study not only demonstrates the material's effectiveness in structural applications but also emphasizes its potential to significantly reduce the carbon footprint of the construction industry, a major contributor to global greenhouse gas emissions. Regarding performance in different environmental conditions, a study conducted in Thailand examined the use of geopolymer concrete in marine environments. This research is particularly relevant given the challenges faced

K. K. Poloju and K. Srinivasu, *Geopolymer Concrete*,
SpringerBriefs in Applied Sciences and Technology,
https://doi.org/10.1007/978-981-96-2479-9_7

by coastal infrastructure due to exposure to aggressive marine conditions. The study found that geopolymer concrete exhibited superior resistance to chloride penetration and sulfate attack compared to ordinary Portland cement concrete.

This enhanced durability makes geopolymer concrete particularly suitable for coastal infrastructure, where traditional concrete often suffers from accelerated deterioration due to the harsh marine environment. The findings of this study suggest that geopolymer concrete could potentially extend the service life of coastal structures, reducing maintenance costs and improving long-term sustainability. Additionally, a case study in the United States explored the use of geopolymer concrete in bridge construction, specifically for precast bridge decks in Virginia.

This application is significant as it demonstrates the material's potential in critical infrastructure projects. The geopolymer concrete used in these bridge decks demonstrated excellent durability and strength properties, even when subjected to freeze–thaw conditions. This performance characteristic is crucial for regions with harsh winter climates, where traditional concrete often suffers from deterioration due to repeated freezing and thawing cycles. The success of this case study suggests the potential for widespread adoption of geopolymer concrete in regions with challenging climate conditions, potentially leading to more durable and longer-lasting infrastructure.

The diverse range of case studies and research findings highlight the adaptability and performance benefits of geopolymer concrete across various construction applications and environmental conditions. From airports and institutional buildings to marine structures and bridges, geopolymer concrete has demonstrated its capability to meet and often exceed the performance of traditional concrete. Its ability to withstand heavy loads, resist chemical attacks, and perform well in harsh environments makes it a versatile material for a wide range of construction projects. Furthermore, the significant reduction in carbon emissions associated with geopolymer concrete production and use aligns well with global efforts to mitigate climate change.

As the construction industry faces increasing pressure to reduce its environmental impact, the adoption of geopolymer concrete could play a crucial role in achieving sustainability goals without compromising structural performance. However, it is important to note that while these case studies and research findings are promising, wider adoption of geopolymer concrete still faces challenges. These include the need for standardization of mix designs, development of comprehensive design codes, and addressing concerns related to long-term performance and durability.

Continued research and additional large-scale implementations will be crucial in overcoming these challenges and establishing geopolymer concrete as a mainstream construction material. In conclusion, the successful implementation of geopolymer concrete in various projects worldwide demonstrates its potential to revolutionize the construction industry. Its superior performance in diverse environmental conditions, coupled with its significantly lower environmental impact, positions geopolymer concrete as a promising solution for sustainable and durable construction in future. As more data becomes available from long-term studies and additional large-scale projects, the confidence in and adoption of geopolymer concrete is likely to increase, potentially leading to its widespread use in the construction industry. These case studies and research findings provide compelling evidence for the potential of geopolymer concrete to revolutionize the construction industry. As more long-term data becomes available and additional large-scale projects are completed, the confidence in and adoption of geopolymer concrete is likely to increase. This could lead to its widespread use in various construction applications, particularly in environmentally sensitive areas or regions with harsh climatic conditions.

7.1 Summary

Geopolymer concrete has been successfully used in various construction projects worldwide, demonstrating its versatility and superior performance compared to traditional concrete. Case studies, such as the Brisbane West Well Camp Airport in Australia and the Global Change Institute building at the University of Queensland, showcase its ability to withstand heavy loads, resist chemical attacks, and significantly reduce carbon emissions.

Research in Thailand and the United States has also shown geopolymer concrete's excellent durability in marine environments and its ability to perform well under freeze–thaw conditions. Despite the promising results, wider adoption of geopolymer concrete still faces challenges, including the need for standardization and the development of comprehensive design codes.

As more data becomes available from long-term studies and large-scale projects, the confidence in and adoption of geopolymer concrete are likely to increase, potentially leading to its widespread use in the construction industry.

7.2 Highlights

1. Geopolymer concrete has been successfully implemented in numerous construction projects globally, demonstrating versatility and efficacy.
2. Brisbane West Well Camp Airport utilized geopolymer concrete, resulting in significant carbon emission reduction.
3. Global Change Institute building at the University of Queensland achieved a 70% carbon emission reduction.
4. Geopolymer concrete exhibited superior resistance to chloride penetration and sulfate attack in marine environments.
5. Precast geopolymer concrete bridge decks in Virginia demonstrated excellent durability and strength properties.
6. Adoption of geopolymer concrete aligns with global efforts to mitigate climate change.
7. Continued research and large-scale implementations are crucial for establishing geopolymer concrete as a mainstream construction material.

Chapter 8
Geopolymer Concrete: Future Directions

This chapter discussed the opportunities for new applications and the requirement for more R&D.

Geopolymer concrete has significant potential as a future material to be used in various industries, for example, in sustainable construction and infrastructure areas. With a focus on sustainability and greener construction coming into play, geopolymer concrete can find a place more frequently in mega projects like bridge construction, highways, and high-rise buildings. This could make it an attractive choice for marine structures, chemical plants, and other facilities that are exposed to corrosive environments because of their potentially improved durability and resistance to harsh conditions.

Elsewhere, as the material's fire resistance characteristics are likely to make it suitable for areas where fires pose a significant threat or for important infrastructure that requires higher levels of fire protection, geopolymer concrete, with its distinct advantages such as a lower carbon footprint compared to conventional Portland cement concrete, has the potential to revolutionize the industry. The ability to recycle industrial by-products like fly ash and slag helps address their waste issues, meaning that instead of being dumped in landfills, they become raw materials for new buildings—contributing to the circular economy.

This is consistent with worldwide indicators to reduce greenhouse gas emissions and implement sustainable development trends. More R&D needs to be done with geopolymer concrete. Some of the main research avenues needed are reduced mix designs based on different applications, enhanced long-term performance and durability, and standardized testing methods as well as QC/QA procedures. Another aspect is to investigate using alternative precursor materials to widen the spectrum of geopolymer formulations and reduce dependency on industrial waste matter. This may involve looking into naturally available materials or other waste streams that could act as suitable raw materials for geopolymer manufacturing. In order to gain a wider industrial report, investigations into the economics and life cycle analysis

K. K. Poloju and K. Srinivasu, *Geopolymer Concrete*,
SpringerBriefs in Applied Sciences and Technology,
https://doi.org/10.1007/978-981-96-2479-9_8

of geopolymer concrete, as compared to the ordinary Portland cement concrete candidates, will also be important.

This involves reviewing the total cost of ownership, which would include filling all required initial material costs, transportation, installation, maintenance, and end-of-life disposal or recycling. The real test will come in three to five years to understand the long-term economic and environmental benefits, as well as the challenges it creates for those stakeholders who would be asked to invest in this novel material. With the maturation of these technologies, more attention should be directed to scaling up production and manufacturing processes, as well as addressing field-level issues such as mix and placement on-site.

It could include designing specific machinery to produce geopolymer concrete or refurbishing existing concrete production plants for a compatible geopolymer mix. Training programs for construction workers and engineers will also be needed to ensure appropriate handling and use of geopolymer concrete on actual projects. Alternative uses for geopolymer concrete go beyond traditional construction-type work. The material has unique properties that make it suitable for special applications like 3D printing of complex architectural elements, prefabricated building components, and smart infrastructure when some sensors are embedded into the concrete. Its high-temperature and attack-resistant attributes also make it possible for the material to be utilized in geothermal energy projects, nuclear waste storage facilities, and other more extreme scenarios.

The advancement of geopolymer concrete technology will certainly require some collaboration between academia, industry, and government agencies. This consists of funding for research, both pilot projects to show its real-world performance and standards/specifications to inform its application in construction. By sharing knowledge and facilitating international collaboration, it is possible for geopolymer concrete to be used much more quickly within certain regions facing infrastructure hurdles or environmental issues. Since the world is gaining more knowledge about geopolymer concrete, investors and entrepreneurs who look for sustainable construction products will become interested in this product. This could cause the development of companies that focus directly on geopolymer production, application, and service. The expansion of this sector might additionally stimulate innovation in supporting areas, including admixtures, reinforcement materials, and building equipment that is predominantly used for geopolymer concrete. Finally, the future of geopolymer concrete seems bright and could play an instrumental role in both revolutionizing the construction industry and helping to meet sustainable development goals.

Achieving this potential will, however, take a continuum of research, investment, and collaboration among different stakeholders in the construction ecosystem. Geopolymer concrete might become part of the future built environment as technology develops and matures, with potentially less environmental impact and providing greater durability than traditional concrete. By using geopolymer concrete, there can be extreme impacts from urban planning to design stages. Geopolymer concrete could play an important role in the creation of sustainable infrastructure and buildings as cities become more sustainable and resilient. We could see it being used more in green building certifications like LEED (Leadership in Energy and Environmental Design), giving projects extra points for using sustainable materials.

The versatility of the material might also allow for other innovative architectural designs that can be difficult, if not impossible, with regular concrete. This may open fresh avenues of creativity for architects and engineers but also provide an opportunity to tackle environmental problems simultaneously. The new look for this middle ground could be urban built forms that reflect the properties of geopolymer concrete—both visually and performatively. Also, in the infrastructure, geopolymer concrete has significant potential to prolong the lifespan of critical assets. Bridges, tunnels, and roads built using this archeological material may require less frequent maintenance and replacement—lowering lifecycle costs and reducing disruption to transportation networks. This durability may be especially important in developing countries that require infrastructure investment as a prerequisite for economic growth and social development.

Another area that deserves attention is the suitability of geopolymer concrete for disaster-resilient construction. Therefore, if widely used, it can be a building material for high-rise buildings in areas that are prone to earthquakes or extreme weather events due to its strong and durable traits. With climate change making natural disasters more common and intense, the resilience that geopolymer concrete can offer may eventually be seen as priceless. Geopolymer Concrete for Global Sustainability Goals in Addition to Sustainable Development Goals (SDGs) by the United Nations. Its ability to reduce carbon emissions meets climate action, whereas its use of industrial by-products supports responsible consumption and production.

Sustainable cities and communities are made possible by the durability of this material along with its ability to reduce the amount of waste that goes into construction. Geopolymer concrete technology could also produce sound economic development. With the global demand for low-carbon economies, the production and application industries pertaining to geopolymer concrete may serve as significant economic sources for green jobs. Such positions may include research and development, manufacturing, construction, and environmental assessment. Get trained: training and education will be important to generalizing the use of geopolymer concrete. This may require universities and technical institutions to vary their curricula with courses on geopolymer technology, so that generations of engineers and construction professionals will be familiar with its properties at both ends of the spectrum.

It will also be critical to provide professional development programs to current practitioners to spread knowledge and best practices. Geopolymer concrete could play a role in shaping policy and building codes. Governments may incentivize or regulate its use in ways akin to existing policies that promote other green building materials and practices. This could range from tax incentives to fast-tracked permits to its use mandates in some forms of public construction. This may not only be a milestone in geopolymer concrete technology but could also open ways for further innovations. For example, adding a new class of reinforcement compatible with geopolymer matrices, where even conventional steel reinforcement may not be needed for some applications.

This would enable us to improve both the sustainability and performance properties of geopolymer concrete structures. Geopolymer concrete has been recognized internationally for the opportunities it presents, as such a change will have knock-on effects in global trade and the flow of material. Nations with large quantities of appropriate precursor materials (e.g., fly ash or metakaolin) available may emerge as significant countries supplying the geopolymer value chain. This could open up new avenues for trade as well as change the dynamics of the global construction materials market. In the distant future, large-scale adoption of geopolymer concrete can play a part in changing urban mining. Over its lifecycle, structures and infrastructure built from geopolymer concrete could be recycled, aligning with principles of a circular economy. This may also help create new recycling technology and processes for geopolymer materials. One more area needing further investigation is the potential of geopolymer concrete for carbon sequestration.

Depending on the formulation, geopolymer concrete can capture CO_2 from the atmosphere and that can make these structures carbon sinks. If this characteristic is consistently exploited and scaled up, it could be a considerable step forward in climate change reduction methods in the built environment. With the maturation of geopolymer concrete technology, it can also be used beyond Earth. It can be made using local materials and has the potential to survive extreme conditions, making it a potentially useful building material for other planets or moons. Geopolymer concrete could also be researched with the express purpose of eventually adapting it for extraterrestrial construction by space agencies and private space exploration companies as part of preliminary long-term goals required for sustainable human living on Mars or possibly the Moon. Widespread adoption will be met with institutional challenges, from a long-standing industry resistance to new market entrants to the development timeframe of facilities initially capitalized as assets on balance sheets that have no immediate demand for environmentally sustainable change.

"Combining both worlds of material innovation and business model disruption is certainly not for the weak;" these challenges will demand collective action from policymakers, industry giants and small businesses alike, researchers, and environmental proponents. Finally, the future of geopolymer concrete closely relates to more general sustainability and technology trends along with global development. Its ability to tackle many challenges faced by the construction industry and address environmental targets makes it a technology that we should all track. It bears immense potential as applied research develops and industry implementation follows suit—Geopolymer concrete.

8.1 Summary

Geopolymer concrete is seen as a promising material for future sustainable construction and infrastructure applications. Vertical ducts are suitable for many projects such as bridges, high-rise buildings, and marine structures due to their unique properties like reduced carbon footprint, durability, and resistance to harsh environments. While there is an urgent need to investigate possible optimum mix designs, durability in field conditions, and standard testing methods. The real development of geopolymer concrete technology requires financial assistance, pilot projects, and standards; therefore, collaboration between academic institutions and the industry sectors, along with government agencies, is needed. Geopolymers could be environmentally friendly components of smart and sustainable urban planning, disaster-resistant buildings, and may even help manage global environmental challenges. Hurdles to mainstream usage include opposition from industries resistant to change and the initial capital expenditure needed for new manufacturing plants.

8.2 Highlights

1. The use of geopolymer concrete has potential as a sustainable alternative to conventional concrete because it can minimize carbon footprint and waste associated with the construction stage.
2. Mixed designs, performance improvement, and standard development; however, more research is needed.
3. Industry-academia-government collaboration needs to be encouraged to achieve technological advancement.
4. Geopolymer concrete may change the way we plan our towns, cities, buildings, and other human needs in an environmentally safe way.
5. The robustness and strength of the material can be useful for construction that is disaster resistant.
6. The use of geopolymer concrete can help achieve overall sustainability targets and create green jobs in the world.
7. Challenges involve pushing against entrenched industries and a high initial price tag.

Chapter 9
Conclusion

1. Geopolymer concrete is a very new and sustainable option compared to Portland cement concrete. It employs industrial waste materials, such as fly ash or blast furnace slag, as the main binder, thus recycling materials that would otherwise be in the form of a waste product. This not only reduces the carbon footprint caused by concrete production but also provides an answer to industrial waste management and severity issues.

2. Consider the environmental advantages of geopolymer concrete: CO_2 emission reductions can be as high as 80% when compared to normal concrete production. The bulk of this decrease in CO_2 is due to a fundamental increase in the carbon footprint of making Portland cement, specifically the production of clinker, which is an energy-intensive process. The use of geopolymer concrete may act as a significant tool in reducing the construction industry's solidary growth impact on global greenhouse gas emissions.

3. Geopolymer concrete has been shown to have significant advantages related to durability, such as very high acid and sulfate resistance and fire resistance. This increased durability is due to the greater stability of the geopolymers, which form a more stable and resistant matrix than cement hydration products. Its higher resistance to chemical attacks makes geopolymer concrete beneficial for use in aggressive environments, such as industrial assets or marine infrastructures.

4. Geopolymer concrete is one of the construction materials with the highest early strength and fast strength development. This quality accelerates the construction process, resulting in possible time and cost savings on a project. This rapid development of heat is extremely advantageous in precast applications, especially when there is a necessity for early pull-out (for formwork) or fast-loading structural parts.

5. As opposed to traditional concrete, geopolymer concrete shows lower drying shrinkage and creep. Effectively, dimensional stability is significantly enhanced,

© The Author(s), under exclusive license to Springer Nature Singapore Pte Ltd. 2025   103
K. K. Poloju and K. Srinivasu, *Geopolymer Concrete*,
SpringerBriefs in Applied Sciences and Technology,
https://doi.org/10.1007/978-981-96-2479-9_9

as this quality is important to ensure the structure remains usable and maintains its integrity over time. It exhibits lower shrinkage and creep, which may cause reduced cracking and diminished long-term deformation of the structural components, thereby increasing the performance and service life of structures, buildings, and infrastructure.

6. As a result, this material has excellent thermal stability and chemical resistance, making it ideal for dry or harsh environments as well as unique applications. Geopolymer concrete does not break down for more months than other concrete types; therefore, it is a great choice for cold or fire-resistant structures or industrial applications where it encounters high temperatures. Due to its chemical resistance, it is still able to resist fire and chemicals that might be present in a wastewater treatment plant or at places dealing with raw chemical storage.

7. Geopolymer concrete generally consumes less energy and uses less water in the manufacturing process compared to regular cement concrete. This means carbon emissions per ton of leather are lower, as well as less pressure on scarce water resources—an important sustainability aspect. This reduction in energy requirements can also result in lower production costs, though these benefits may be offset by the comparatively higher cost of some geopolymer precursors.

8. Geopolymer concrete has many benefits, but it is also facing some challenges. One of its main drawbacks is that many applications require heat curing to develop maximum strength and durability. This necessity can be a drawback, limiting on-site use and raising the complexity and prices of manufacturing precast components. Finally, the absence of long-term performance data and standardized mix design protocols hinders widespread adoption within the construction sector. This leads to uncertainty on the part of engineers and contractors who might err on the side of caution and refuse to specify or adopt a material that lacks a significant history of performance data.

9. However, current limitations in the applicability of geopolymer concrete have attracted ongoing research in some key areas for improvement. Some areas include optimizing mix design to improve performance while lowering costs, finding superior ambient curing applications to negate thermal curing, and establishing proper standards/guidelines on mix design/preparation and quality control. Other research is examining the long-term strength and service life of geopolymer concrete structures to dispel safety concerns regarding these materials for critical infrastructure projects.

10. There is a bright future for the use of geopolymer concrete in different fields, as this specialized form of concrete can be beneficial compared to traditional methods. This makes it highly applicable for precast pieces, as controlled environments can ensure that curing and quality control processes are optimal. Their resistance to chloride ingress and sulfate attack makes them suitable for application in marine structures, which could help prolong the service life of coastal infrastructure. Potential uses also include chemical storage facilities, structures that require high-temperature-resistant materials, and fast-curing materials for

use in rehabilitation of infrastructure. While research moves forward and experience is gained with applications, the assortment of uses for geopolymer concrete is expected to continue expanding.

Bibliography

A. Aleem, P. Arumairaj, Geopolymer concrete—a review. Int. J. Eng. Sci. Emerg. Technol. **1**, 118–122 (2012)

M.A.M. Ariffin, M.A.R. Bhutta, M.W. Hussin, M.M. Tahir, N. Aziah, Sulfuric acid resistance of blended ash geopolymer concrete. Constr. Build. Mater. **43**, 80–86 (2013)

T. Bakharev, The durability of geopolymer materials in sodium and magnesium sulphate solutions. Cem. Concr. Res. **35**(6), 1233–1246 (2005)

R.R. Bellum, K. Muniraj, S.R.C. Madduru, Influence of activator solution on microstructural and mechanical properties of geopolymer concrete. Materialia **10**, 100659 (2020)

J. Davidovits, *Geopolymer Cement a Review 2013* (2013)

P.S. Deb, P.K. Nath, P.K. Sarker, Properties of fly ash and slag blended geopolymer concrete cured at ambient temperature, in *ISEC 2013* (2013)

A. Fernández-Jiménez, J.Y. Pastor, A. Martín, A. Palomo, High-temperature resistance in alkali-activated cement. J. Am. Ceram. Soc. **93**(10), 3411–3417 (2010)

IS 2386 (Part-1), *Particle Size and Shape* (BIS, New Delhi, 1963)

IS 4031 (Part-1), *Methods of Physical Tests for Hydraulic Cement* (BIS, New Delhi, 1996)

U.M. Jumppanen, U. Diederichs, K. Hinrichsmeyer, *Material Properties of F-Concrete at High Temperatures* (1986)

S. Kumaravel, K. Girija, Acid and salt resistance of geopolymer concrete with varying concentrations of NaOH. J. Emerg. Technol. Innov. Res. **4**, 1–3 (2013)

J. Lehne, P. Felix, *Making Concrete Change Innovation in Low-Carbon Cement and Concrete* (Chatham House, 2018)

G. Mallikarjuna Rao, Effect of fly ash and GGBS combination on mechanical and durability properties of GPC. Adv. Concr. Constr. **5**(4), 313–330 (2017). https://doi.org/10.12989/ACC.2017.5.4.313

G. Mallikarjuna Rao, T.G. Rao, Final setting time and compressive strength of FAG-based geopolymer paste and mortar. Arab. J. Sci. Eng. **40**, 3067–3074 (2016)

G. Mallikarjuna Rao, K.K. Sathish, K.K. Poloju, K. Srinivasu, *Emphasis on Geopolymer Concrete with Single Activator and Conventional Concrete with Recycled Aggregate and Data Analysis Using Artificial Neural Networks* (Advances in Mechanical Sciences, Hyderabad, India)

R.K. Manchiryal, K.K. Poloju, Y.A.A. Al Balushi, W.N.M. Al Banna, Variation of sodium hydroxide concentration impacts the rheological properties of geopolymer paste. Int. J. Adv. Appl. Sci. **10**(1), 62–68 (2023)

G. Mohankumar, R. Manickavasagam, Study on the development of class C flyash based GPC by ambient curing. Int. J. Appl. Eng. Res. **12**, 1227–1231 (2017)

M.S. Morsy, S.H. Alsayed, Y. Al-Salloum, T. Almusallam, Effect of sodium silicate to sodium hydroxide ratios on strength and microstructure of fly ash geopolymer binder. Arab. J. Sci. Eng. **39**, 4333–4339 (2014)

F.N. Okoye, S. Prakash, N.B. Singh, Durability of fly ash based geopolymer concrete in the presence of silica fume. J. Clean. Prod. **149**, 1062–1067 (2017)

S. Oyebisi, A. Ede, O. Ofuyatan, T. Alayande, G. Mark, J. Jolayemi, S. Ayegbo, Effects of 12 molar concentrations of sodium hydroxide on the compressive strength of geopolymer concrete. IOP Conf. Ser. Mater. Sci. Eng. **413**, 012066 (2018)

K.K. Poloju, K. Srinivasu, Impact of GGBS and strength ratio on mechanical properties of geopolymer concrete under ambient curing and oven curing. Mater. Today Proc. **42**(Part 2), 962–968 (2021). ISSN 2214-7853

K.K. Poloju, K. Srinivasu, G. Mallikarjuna Rao, Study on mechanical characterization of geopolymer cement mortar with single solution and combined solution. J. Xi'an Univ. Arch. Technol. **XII**(VIII), 481–487 (2020)

K.K. Poloju, K. Srinivasu, T.V.S. Vara Laxmi, G. Mallikarjuna Rao, Method of determining characteristics of geopolymer concrete under elevated temperatures. NeuroQuantology **20**(10), 11063–11071 (2022)

K.K. Poloju, S. Annadurai, R.K. Manchiryal, M.R. Goriparthi, P. Baskar, M. Prabakaran, J. Kim, Analysis of rheological characteristic studies of fly-ash-based geopolymer concrete. Buildings **13**(3), 811 (2023)

A.O. Purdon, The action of alkalis on blast-furnace slag. J. Soc. Chem. Ind. **59**(9), 191–202 (1940)

N.P. Rajamane, M.C. Nataraja, J.K. Dattatreya, N. Lakshmanan, D. Sabitha, Sulphate resistance and eco-friendliness of geopolymer concrete. Indian Concr. J. **86**(1), 13 (2012)

U. Rattanasak, K. Pankhet, P. Chindaprasirt, Effects of chemical admixtures on the properties of high-calcium fly ash geopolymers. Int. J. Miner. Metall. Mater. **18**(3), 364–369 (2011)

A.K. Saha, Effect of class F fly ash on the durability properties of concrete. Sustain. Environ. Res. **28**, 25–31 (2018)

S.H. Sanni, R. Khadiranaikar, Performance of geopolymer concrete under severe environmental conditions. Int. J. Civ. Struct. Eng. **3**(2), 396–407 (2012)

S.E. Wallah, D. Hardjito, D.M.J. Sumajouw, B.V. Rangan, Sulfate and acid resistance of fly ash-based geopolymer concrete, in *The Australian Structural Engineering Conference (ASEC)* (2005a)

W. Wang, A. Shen, Z. Lyu, Z. He, K.T.Q. Nguyen, Fresh and rheological characteristics of fibre reinforced concrete—a review. Constr. Build. Mater. **296**, 123734 (2021)

O.K. Wattimena, A.D. Hardjito, A review of the effect of fly ash characteristics and their variations on the synthesis of fly ash based geopolymer. AIP Conf. Proc. (2017)

J. Xie, J. Wang, R. Rao, C. Wang, C. Fang, Effects of combined usage of GGBS and fly ash on workability and mechanical properties of alkali activated geopolymer concrete with recycled aggregate. Compos. Part B Eng. **164**, 179–190 (2019)